Singular perturbation techniques applied to integro-differential equations

H Grabmüller
Technische Hochschule Darmstadt

Singular perturbation techniques applied to integro-differential equations

Pitman
LONDON · SAN FRANCISCO · MELBOURNE

PITMAN PUBLISHING LIMITED
39 Parker Street, London WC2B 5PB

FEARON-PITMAN PUBLISHERS INC.
6 Davis Drive, Belmont, California 94002, USA

Associated Companies
Copp Clark Ltd, Toronto
Pitman Publishing New Zealand Ltd, Wellington
Pitman Publishing Pty Ltd, Melbourne

First published 1978

AMS Subject Classifications: (main) 45K05, 45L05, 47A55, 47B35, 65R05
(subsidiary) 45E10, 47G05, 73B30

Reproduced and printed by photolithography
in Great Britain at Biddles of Guildford

ISBN 0 273 08409 7

Preface

This research note is a study of singular perturbation techniques applied to a particular class of linear partial integro-differential equations having kernel functions of convolution type on a semiaxis. The main purposes of this study can be seen in two respects. The first one is the development of the analytic apparatus which necessarily arises from the application of the method of matched asymptotic expansions to the underlying boundary layer problems. The second one is the extension of the classical Wiener-Hopf techniques to integral equations with operator-valued kernels.

Both items, namely singular perturbations and Wiener-Hopf techniques have been used in applied mathematics, and various surveys of results can be found in the existing literature. No such survey is attempted here; the aim of this treatment is to discover a new approach to the solution of initial boundary value problems as they appear for instance in the theory of generalized heat conduction in materials with memory.

The research note was motivated by Professor G.C. Hsiao's lectures on singular perturbation theory which he gave during his stay in Darmstadt in the academic year 1975/76. I am indebted to him for many fruitful discussions. I also wish to express my appreciation to Professor W. Wendland for his encouragement throughout the development of the work and for his help during the process of publishing. Furthermore, I am most grateful to Professors E. Meister, R.P. Gilbert, and A. Jeffrey for their supports on this book. Finally, I would like to express my gratitudes to Frau U. Abou El-Seoud and to Frau T. Ridder who both did an excellent job in typing the manuscript.

H.G.

Contents

1 Introduction

Let Ω denote the open interval $(o,1)$. We deal here with an initial boundary value problem for a linear partial integro-differential equation of the following type:

$$\left.\begin{array}{lll} M_\varepsilon[u] = \varepsilon r(x,t) & \text{for} & (x,t) \in \Omega \times (o,\infty), \\ u(j,t) = g_j(t) & \text{for} & t \geq o \text{ and } j = o,1, \\ u(x,o) = h(x) & \text{for} & x \in \Omega. \end{array}\right\} \qquad (P_\varepsilon)$$

Here the operator M_ε is defined by

$$M_\varepsilon[u] \equiv (k + \partial/\partial t)u$$
$$+ \ \varepsilon\{c\ \partial^2 u/\partial x^2 + \int_o^\infty [k_o(t-s)\ \partial^2/\partial x^2 + k_1(t-s)(k+\partial/\partial s)]u(x,s)ds\},$$

where ε is a small positive parameter and where $k > o$ and $c \neq o$ are constants independent of ε. Unless otherwise stated, the kernel functions $k_j (j=o,1)$ are assumed to belong to the space $L^1(\mathbb{R})$, i. e. the space of complex valued functions which are defined on the reals and which are integrable in the sense of Lebesgue.

The main purpose of this paper is to give an asymptotic expansion for the exact solution of the problem (P_ε) which is valid in some sense as $\varepsilon \to o+$. The present method generalizes the one used in our paper [23] where we have discussed existence of a zeroth order approximation for the solution of (P_ε). Moreover, the proof in [23] was accomplished by considering the problem as an abstract equation in the Hilbert space $L^2(o,\infty)$. The resulting expansion was valid only in the L^2 norm. In this paper, we treat the full problem (P_ε), and our method of proof allows us to use $L^p(o,\infty)$ norms $(1 \leq p \leq \infty)$ as well as uniform norms everywhere. We also obtain existence and uniqueness of the exact solution.

The problem (P_ε) belongs to the class of so called singular perturbation problems. Problems of this nature appear in many branches of applied mathematics, for instance in celestial mechanics, oscillation and wave theory, electromagnetic theory, oceanography, chemical reactor theory, viscous flow theory. They have received considerable attention in recent years, and a vast multitude of research papers as well as some monographs and books are devoted to them. The reader will become best oriented on this field by a recent survey of A.Erdélyi [17] as well as by the older surveys of R. P. Kanwal [28] and P. A. Lagerstrom [35].

Problems of the particular type (P_ε) are of considerable interest in the theory of generalized heat conduction in materials with fading memory. Some general theories of this nature have been developed recently by M. E. Gurtin and A. C. Pipkin [24], J. Meixner [36], J. W. Nunziato [38], and others. In this context, the parameter ε is related to the coefficient of heat capacity, and the limiting case $\varepsilon \to o+$ represents a conductor with an infinite heat capacity. For further informations of the physical background the reader is also referred to R. K. Miller [37] and to the author's papers [21,22]. A problem similar to (P_ε) also arose in the theory of linear viscoelasticity where the history of the elastic material involves integral operators of the type M_ε. We refer to C. M. Dafermos [10,11,12] and M. Slemrod [43].

Perturbation problems are, broadly speaking, problems (P_ε) depending on a small parameter ε. If the solution, say u_ε, of such a problem depends analytically on ε, then it can be represented formally by the Taylor expansion about $\varepsilon = o$, where the zeroth order term u_o is the solution of the limiting problem (P_o). Such a representation to a finite number of terms with a remainder of appropriate order holds in many cases when u_ε is not an analytic function of ε. A singular perturbation problem is one in which this simple expansion fails. However, the reasons for such a failure can be of quite different nature, and they are specific for the particular problem. On this account, there is no acceptable general definition of singular perturbations, and there cannot be a general theory.

In our particular case, the problem (P_ε) is singular in the sense that the initial and boundary conditions prescribed for (P_ε) are not appropriate to the limiting equation which is obtained by passing to the limit $\varepsilon \to o+$. Since this equation constitutes an ordinary differential equation of first order it is impossible to satisfy the prescribed boundary conditions in general.

From experiences it is suggested that for the moment the boundary conditions should be ignored. A trial solution

$$U(x,t;\varepsilon) = \sum_\mu \varepsilon^\mu U_\mu(x,t) \tag{1.1}$$

leads, upon substitution into the differential equation and comparison of coefficients of like powers of ε, together with the initial condition to a set of well posed initial value problems for the determination of U_μ. By solving these problems successively one arrives at the <u>outer expansion</u> (1.1) which is a possibly divergent series. However, a comparison of the partial sum of m-th order with the exact solution (which is understood in the classical sense and which shall be assumed to exist for the present) shows that the difference is of order ε^{m+1} uniformly for x outside some boundary layer region. Of course at $x = o$ and $x = 1$ the outer expansion fails to approximate the solution.

The boundary layer is investigated by means of a stretching transformation $\tilde{x}_o = x/\sqrt{\varepsilon}$ (respectively $\tilde{x}_1 = (1-x)/\sqrt{\varepsilon}$) where $\tilde{x}_j$ $(j = o,1)$ are the new independent variables. The inner expansions

$$V^j(x,t;\varepsilon) = \sum \varepsilon^{\mu} V^j_{\mu}(\tilde{x}_j,t) \qquad (j = o,1) \tag{1.2}$$

lead to a set of initial boundary value problems where either the boundary condition at $x = o$ (i.e. $\tilde{x}_o = o$) or the boundary condition at $x = 1$ (i.e. $\tilde{x}_1 = o$) can be satisfied, and where it is opportune to require a homogeneous initial condition. However, in both cases the respective second boundary condition makes no sense. Now the question arises, what is the proper additional condition which should be required to obtain a set of well posed initial boundary value problems? Again, a sufficient answer can be received from experiences. In our approach, the proper condition is a matching condition which was first proposed by L. Prandtl [41] and which was successfully used by M. I. Višik and L. A. Lyusternik [48], W. Eckhaus and E. M. de Jager [16], J. Cole [9], M. Van Dyke [14] and L. E. Fraenkel [18]. Matching is one of the most important techniques in singular perturbations and it can be expressed in various forms. In our context, matching has to be understood in the sense that the boundary correction terms $V^j_{\mu}(\tilde{x}_j,t)$ should tend to zero as $\tilde{x}_j$ tends to infinity.

An essential role in our investigations is played by the theory of convolution operators. Some general outlines of this theory (which are appropriate to our approach) have been developed by M. G. Krein [34] and E. Gerlach [20].

To be precise, a convolution operator is to be understood as an integral operator on a half-line having a kernel function which depends upon the difference of the arguments. We shall show that the inverse G of the particular differential operator $k + \partial/\partial t$ is a convolution operator acting on some suitable function spaces E. We shall thoroughly study the convolution operators K_μ ($\mu = o,1$) which are associated to the kernel functions k_μ. In particular, we will see that the operator $A = G(cI + K_o)$ is basic for the construction of the inner expansion (1.2) which is accomplished by using the technique of Dunford-Taylor integrals.

The existence proofs of the respective Dunford-Taylor integrals are one of the crucial businesses of this paper. To guarantee existence we require the <u>Sector condition</u> which is precised in Section 2.3. This condition restricts the values of the constant c and of the $L^1(\mathbb{R})$ norm of k_o such that the <u>reduced resolvent</u> $(\lambda A-I)^{-1}$ exists as a continuous linear operator for all λ in a sector $|\arg \lambda| \leqq \theta$, $\theta > o$, and such that the operator norm $\|(\lambda A-I)^{-1}\|$ is bounded uniformly for all λ.

Clearly, one requires also certain smoothness properties for the given data, i.e. for the functions r, g_j ($j = o,1$) and h. The <u>Compatibility conditions</u>: $h(j) = g_j(o)$ ($j = o,1$), are quite natural. They guarantee that the initial and boundary data are well matched at the corners of the rectangle $[o,1] \times [o,\infty)$.

In order to secure the existence of the outer expansion (1.1) - at least of its partial sum of m-th order - we need an additional <u>Regularity condition</u>. Precisely, the initial function $h(x)$ is required to belong to the space $C^{2m+2+\alpha}[o,1]$ for some $o < \alpha < 1$, whereas the right-hand side $r(x,t)$ is required to be a function of $C^{2m+\alpha}([o,1];E)$. The Banach space E can be realized by any space $L^p(o,\infty)$ $(1 \leq p \leq \infty)$ as well as by a suitable space of continuous functions. In any case it will be required that the boundary data g_j $(j = o,1)$ also belong to E.

Under the above conditions, the main result of this paper can be summarized in the following theorem.

THEOREM (1.A). Let the Sector condition and the Compatibility conditions be satisfied, and suppose that the Regularity condition holds. Assume that $E \subset L^\infty(o,\infty)$ is a suitable space of continuous functions, and let $k_1 \in L^1(\mathbb{R}) \cap L^\infty_{loc}(o,\infty)$. Then for each small $\varepsilon > o$ the problem (P_ε) has a unique solution $u \in C^2((o,1);E) \cap C([o,1];E)$ which can be represented by

$$u(x,t;\varepsilon) = \bar{u}(x,t;\varepsilon) + Z(x,t;\varepsilon) \qquad (o \leq x \leq 1, \quad t \geq o) .$$

Here

$$\bar{u}(x,t;\varepsilon) = U(x,t;\varepsilon) + V^o(x,t;\varepsilon) + V^1(x,t;\varepsilon)$$

is an m-th order approximate solution to (P_ε) where the functions U, V^o and V^1 are defined by the m-th order partial sums of the expansions (1.1) and (1.2) respectively. The functions V^j $(j = o,1)$ are boundary layer terms having the properties, that for any fixed $\eta \in (o,1)$

$$\max_{\eta \leq x \leq 1} ||V^o(x,\cdot;\varepsilon)||_E \to o \quad \text{and} \quad \max_{o \leq x \leq 1-\eta} ||V^1(x,\cdot;\varepsilon)||_E \to o \quad \text{as} \quad \varepsilon \to o+,$$

with an exponential rate of decay. Moreover, the remainder function Z is small of order ε^{m+1} in the sense that

$$\max_{0 \leq x \leq 1} ||Z(x,\cdot;\varepsilon)||_E = O(\varepsilon^{m+1}) \quad \text{as} \quad \varepsilon \to 0+ .$$

The assertions hold for the case where $E = L^p(0,\infty)$ $(1 \leq p \leq \infty)$, if in addition k_1 is an element of the space $L^q(-\infty,0)$ $(p^{-1}+q^{-1} = 1)$.

The proof of this theorem is given in Section 6.3 and uses an existence result for the determination of the remainder function Z. This result is established in Sections 6.1 and 6.2. Chapter 5 is concerned with the determination of the boundary layer, i.e. the inner expansion (1.2), whereas the outer expansion (1.1) is developed in chapter 4. Chapter 2 contains preliminary material on convolution operators and their resolvent functions. Chapter 3 is concerned with certain operator families which generate the solutions of the boundary layer problems.

More general problems than the one studied here can be treated by variants of the methods of this paper, e.g. problems where the given data h, g_j $(j = 0,1)$ and r depend upon the parameter ε and where the respective Taylor expansions about $\varepsilon = 0$ exist up to the order m. Moreover, our results can be also extended to an arbitrary bounded domain Ω in n-dimensional Euclidean space provided that it has a sufficient smooth boundary. In Section 2.6 there are contained even some aspects of possible generalizations to non-linear problems.

Throughout this paper, we shall use the letter C in order to denote various constants which are not necessary always equal.

2 Resolvents of convolution operators and of related operators

The purpose of this chapter is to present some fundamental results on the resolvent function associated with convolution operators. The theory of these operators, considered in a broad variety of function spaces, constitute the core of the first section. This section should be understoood as a brief report upon well-known facts. Moreover, the reader is made familiar with the minimum of notations used in this paper.

More special operators which are closely related to the problem under consideration are introduced in the second section. For these operators we derive a characterization of the resolvent set in terms of purely algebraic conditions which arise from Fourier transform technique.

In our approach an important role is played by operators whose resolvent sets contain a sector of the complex plane with vertex at the origin. The fundamental hypothesis (H1) which is introduced in Section 2.3, expresses a sufficient condition for the validity of this 'sector'-property.

The next two sections are devoted to the establishment of several boundedness results on the resolvent which will be needed in the subsequent analysis.

In the last section we reconsider certain familiarities of proof techniques used in the preceding parts to monotonicity methods involving accretive operators in a Banach space. It turns out that monotonicity methods are the link between the linear theory and a possible generalization to nonlinear problems.

2.1. CONVOLUTION OPERATORS ON SPECIAL BANACH SPACES

As usual, let us denote by $\mathbb{R}$ ($\mathbb{C}$) the set of real (complex) numbers and by $\mathbb{N}$

the set of natural numbers. The symbols $\mathbb{R}_+$ and $\bar{\mathbb{R}}_+$ will be used for the semi-infinite intervals $(o,+\infty)$ and $[o,+\infty)$. We shall use the notations $\partial_t := \partial/\partial t$ and $\partial_x := \partial/\partial x$.

Following Krein [34, p. 199] we shall introduce a set of spaces of complex-valued functions in which the subsequent analysis can be performed simultaneously.

For $1 \leq p < \infty$, let L^p_+ denote the usual Banach spaces $L^p(\mathbb{R}_+)$ of measurable functions v defined almost everywhere (a.e.) on $\mathbb{R}_+$ and having a finite norm

$$\|v\|_p := \left(\int_o^\infty |v(t)|^p dt \right)^{1/p} .$$

Similarly, L^∞_+ will denote the Banach space $L^\infty(\mathbb{R}_+)$ of measurable functions v defined a.e. on $\mathbb{R}_+$ and having a finite norm

$$\|v\|_\infty := \operatorname{ess.\,sup}_{t \geq o} |v(t)| .$$

The symbols M^c_+ and M^u_+ will denote the spaces of all continuous and bounded functions $v : \bar{\mathbb{R}}_+ \to \mathbb{C}$, and of all uniformly continuous and bounded functions respectively.

Furthermore, C_+ will denote the space of all continuous functions $v : \bar{\mathbb{R}}_+ \to \mathbb{C}$ for which $v(\infty) := \lim_{t \to +\infty} v(t)$ exists, and C^o_+ is the space of all functions $v \in C_+$ for which $v(\infty) = o$.

The spaces M^c_+, M^u_+, C_+ and C^o_+, equipped with the norm $\|\cdot\|_\infty$, all are closed subspaces of L^∞_+ satisfying $C^o_+ \subset C_+ \subset M^u_+ \subset M^c_+ \subset L^\infty_+$.

For simplicity, by $\mathcal{E}$ let us denote the collection of the above spaces:

$$\mathcal{E} := \{L^p_+ \ (1 \leq p \leq \infty),\ M^c_+,\ M^u_+,\ C_+,\ C^o_+\} ,$$

and by $E_\cap$ their intersection, $E_\cap := \bigcap_{E \in \mathcal{E}} E = L^1_+ \cap C^o_+$.

In this work, the <u>convolution operators</u> acting on the spaces $E \in \mathcal{E}$ play a central role. As usual, the convolution operator $K : E \to E$ is a linear integral operator defined by

$$Kv(t) := \int_0^\infty k(t-s)v(s)ds \qquad (t \geq o) ,$$

where $k \in L^1(\mathbb{R})$ is a given complex valued function.

Operators of this type have been the subject of studies by many authors. More special, the most important results in view of our applications were obtained by M. G. Krein [34] and E. Gerlach [20]. Some of their results concerning the solvability of equations of the form $(I+K)v = f$ shall be summarized in the following.

Let $B(E)$ denote the Banach algebra of bounded linear operators from E to E.

Since the Fourier transform

$$\hat{k}(\xi) := \int_{-\infty}^\infty e^{i\xi s}\, k(s)ds \qquad (\xi \in \mathbb{R})$$

is well defined for a function $k \in L^1(\mathbb{R})$, we shall assume that the condition $1 + \hat{k}(\xi) \neq o$ $(\xi \in \mathbb{R})$ can be satisfied. In this case the number

$$w(I+K) := -(2\pi)^{-1} \int_{-\infty}^\infty d_\xi \arg[1+\hat{k}(\xi)] \qquad (2.1)$$

is well defined and is called the <u>winding number</u> of the complex-valued function $1 + \hat{k}(\xi)$. This integer has the meaning of a Fredholm-index associated with the Fredholm operator $I + K$. This shows the following theorem on the invertibility of $I + K$.

THEOREM (2.A). Let $k \in L^1(\mathbb{R})$, and let $E \in \mathcal{E}$. Then we have

(i) $K \in B(E)$ with a bound $\|K\| \leq \|k\|_1$ $(:= \|k\|_{L^1(\mathbb{R})})$.

(ii) $I + K\colon E \to E$ is a homeomorphism if and only if the two conditions

$$1 + \hat{k}(\xi) \neq 0 \quad (\xi \in \mathbb{R}) \quad \text{and} \quad w(I + K) = 0 \tag{2.2}$$

are satisfied.

The proof of this theorem is due to M.G. Krein [34] and shall be omitted here.

We note that the operator $I + K$ has an inverse $(I + K)^{-1} \in B(E)$, if the conditions (2.2) are true. We shall explain that $(I + K)^{-1}$ can also be expressed in terms of certain convolution operators. For this purpose we may introduce the two complex-valued functions $h_{\pm} \in L^1(\mathbb{R})$, defined by the unique solutions of the integral equations

$$\left.\begin{aligned} h_+(t) &= -\int_0^\infty k(t-s)h_+(s)ds - k(t) && (0 \leq t < \infty) \\ h_+(t) &= 0 && (t < 0) \end{aligned}\right\} \tag{2.3}$$

and

$$\left.\begin{aligned} h_-(t) &= -\int_{-\infty}^0 k(t-s)h_-(s)ds - k(t) && (-\infty < t \leq 0) \\ h_-(t) &= 0 && (t > 0). \end{aligned}\right\} \tag{2.4}$$

Then we may denote by $H_{\pm}$ the corresponding convolution operators. Let

$\Gamma \in B(E)$ be defined by

$$\Gamma := H_+ + H_- + H_+H_-, \tag{2.5}$$

then it turns out that $(I + K)^{-1} = I + \Gamma$. Thus the kernel function

$$\gamma(t,s) := h_+(t-s) + h_-(t-s) + \int_0^\infty h_+(t-r)h_-(r-s)dr \tag{2.6}$$

generates the resolvent of the operator $I + K$.

REMARK. In general, the operator Γ is not a convolution operator as can be seen from the representation (2.6) of its kernel. Nevertheless, if $k(t) = o$ for $t < o(t > o)$, the operator Γ reduces to the convolution operator $H_+(H_-)$. Both cases may appear in applications.

2.2. THE RESOLVENT SET $\rho(A^{-1})$

In this section E will denote any of the Banach spaces of the collection $\mathcal{E}$ introduced in the preceding section. The norm of E will be denoted by $\|\cdot\|$. Furthermore, E_o^1 will be the class of <u>locally absolutely continuous</u> functions $v \in E$ with the properties $v(o) = o$ and $\partial_t v \in E$. Clearly, E_o^1 is a linear submanifold of E. We shall always assume that E_o^1 is equipped with the norm topology induced by E.

Next, we consider the linear operator $G^{-1} = kI + \partial_t$, defined for a fixed $k > o$ on the domain $D(G^{-1}) := E_o^1$. We note that G^{-1} is a closed operator mapping bijectively from E_o^1 onto E.

Thus, by the Closed-Graph theorem, the inverse $G := (G^{-1})^{-1}$ exists as a bounded linear operator having the representation

$$Gv(t) = \int_0^t e^{-k(t-s)} v(s)ds \quad (t \geq o). \tag{2.7}$$

Define $g(t) := e^{-kt}$ $(t \geq o)$ and $g(t) = o$ $(t < o)$, then obviously $g \in L^1(\mathbb{R})$, and hence the statements of Theorem (2.A) apply to the convolution operator G.

Given a fixed constant $c \in \mathbb{C}$ and a complex-valued function $k_o \in L^1(\mathbb{R})$, we denote by K_c the convolution operator

$$K_c v(t) := (cI + K_o)v(t) = cv(t) + \int_0^\infty k_o(t-s)v(s)ds \quad (t \geq o).$$

It is immediately obvious that $K_c^{-1} \in B(E)$ exists if and only if

$$c + \hat{k}_o(\xi) \neq o \quad (\xi \in \mathbb{R}) \quad \text{and}$$

$$w(K_c) = -(2\pi)^{-1} \int_{-\infty}^{\infty} d_\xi \arg[c + \hat{k}_o(\xi)] = o. \tag{2.8}$$

We shall always assume that these conditions are satisfied.

Finally, we introduce the closed bijective linear operator $A^{-1} := K_c^{-1} G^{-1} : E_o^1 \to E$, and its bounded inverse $A = GK_c \in B(E)$. It turns out that the <u>resolvent</u> $R(\lambda; A^{-1}) := (\lambda I - A^{-1})^{-1}$ of A^{-1} can be used for the construction of solutions of the problem (P_ε). Therefore, a rigorous study of the resolvent is required including a precise determination of the <u>resolvent set</u> $\rho(A^{-1})$ and a detailed analysis of $\|R(\lambda; A^{-1})\|$ as $|\lambda| \to \infty$.

Subsequently we shall establish necessary and sufficient conditions for a number $\lambda \in \mathbb{C}$ to belong to $\rho(A^{-1})$. To this purpose let $\hat{h}: \mathbb{R} \to \mathbb{C}$ be defined by

$$\hat{h}(\xi) := (c+\hat{k}_o(\xi)) \quad (k-i\xi)^{-1} \quad (\xi \in \mathbb{R}). \tag{2.9}$$

Then we shall prove that

$$\lambda \in \rho(A^{-1}) \iff \begin{cases} \text{(i)} \quad \lambda\hat{h}(\xi)-1 \neq o \quad (\xi \in \mathbb{R}), \\ \\ \text{(ii)} \quad -(2\pi)^{-1} \int_{-\infty}^{\infty} d_\xi \arg[\lambda\hat{h}(\xi)-1] = o. \end{cases} \tag{2.10}$$

To begin with, we will give an equivalent representation of the resolvent $R(\lambda;A^{-1})$.

PROPOSITION (2.B). For any $\lambda \in \rho(A^{-1})$, we have $A(\lambda A-I)^{-1} = R(\lambda;A^{-1}) = (\lambda A-I)^{-1}A$.

<u>Proof.</u> Let $\lambda \in \rho(A^{-1})$. Then $\lambda I-A^{-1}: E_o^1 \to E$ is a bijection. Multiplication by $A: E \to E_o^1$ from the right yields a homeomorphism $(\lambda I-A^{-1})A = \lambda A-I: E \to E$, and hence it follows that $A^{-1}(\lambda I-A^{-1})^{-1} = (\lambda A-I)^{-1}$. Thus $A(\lambda A-I)^{-1} = R(\lambda;A^{-1})$. Now, starting from the identity $A(\lambda I-A^{-1}) = \lambda A-I: E_o^1 \to E_o^1$, it is immediately obvious that $R(\lambda;A^{-1}) = (\lambda A-I)^{-1}A$.

We shall call the operator $(\lambda A-I)^{-1}$ the <u>reduced</u> <u>resolvent</u> of A^{-1}, and we shall denote it by $R'(\lambda;A)$. Clearly, if $\lambda \in \rho(A^{-1})$, then $R'(\lambda;A): E \to E$ is a homeomorphism. We may ask whether the converse is also true.

In fact, let $R'(\lambda;A): E \to E$ be a homeomorphism. For any $v \in E_o^1$ let $w := R'(\lambda;A)v$. Then $\lambda Aw = w + v \in E_o^1$, and hence $w \in E_o^1$ because E_o^1 is a linear manifold. Conversely, let $w \in E_o^1$ and $v \in E$ such that $w = R'(\lambda;A)v$. Again, $\lambda Aw = w + v \in E_o^1$, and hence $v \in E_o^1$. Thus $R'(\lambda;A)$ is a bijection from E_o^1 onto E_o^1. Consequently, the inverse operator $(R'(\lambda;A)A)^{-1} = \lambda I - A^{-1}: E_o^1 \to E$ exists. In particular, we have proved that

$$\lambda \in \rho(A^{-1}) \iff R'(\lambda;A): E \to E \quad \text{isomorphic}. \tag{2.11}$$

At a first glance, we may be tempted to deduce the desired characterization (2.10) by an application of Theorem (2.A) to the above result. Unfortunately, $A = GK_c$ is not a convolution operator, and hence Theorem (2.A) is not directly applicable. However, A can be decomposed into the sum of a convolution operator and a compact operator.

LEMMA (2.C). The operator $A = GK_c$ admits a decomposition in the form $A = H-V$, where $H \in B(E)$ is a convolution operator generated by the $L^1(\mathbb{R})$-kernel.

$$h(t) := cg(t) + (g*k_o)(t) = cg(t) + \int_{-\infty}^{\infty} g(t-r)k_o(r)dr,$$

and where V is a one-dimensional (and hence compact) operator defined by

$$Vv(t) := e^{-kt}(K_oGv)(o+) \quad (t \geq o).$$

<u>Proof.</u> Let $\check{k}_o(t) := k_o(-t)$ $(t \in \mathbb{R})$. Then we have

$$\begin{aligned} GK_cv(t) &= c\int_o^{\infty} g(t-s)v(s)ds + \int_o^{\infty} g(t-s)\int_o^{\infty} k_o(s-r)v(r)drds \\ &= c\int_o^{\infty} g(t-s)v(s)ds + \int_o^{\infty} v(s)\int_{-\infty}^{\infty} g(t-s-r)k_o(r)drds \\ &\quad - \int_o^{\infty}\int_o^{\infty} \check{k}_o(s+r)g(t+r)drv(s)ds \end{aligned}$$

for every $v \in E$, where we have made free use of Fubini's theorem. Since $g(t) = e^{-kt} (t \geq o)$ we conclude that

$$\int_0^\infty \int_0^\infty \check{k}_o(s+r)g(t+r)dr\, v(s)ds = e^{-kt} \int_0^\infty e^{ks} v(s) \int_s^\infty e^{-kr}\, \check{k}_o(r)drds$$

$$= e^{-kt} \int_0^\infty k_o(-s) \int_0^s e^{-k(s-r)}\, v(r)drds = e^{-kt} (K_o Gv)(o+).$$

Thus the proof is complete.

We are now in a position to prove the main result of this section, namely the implications (2.10). As a byproduct of the proof we will obtain a remarkable representation of the reduced resolvent $R'(\lambda;A)$ in terms of the operator $(\lambda H-I)^{-1}$. Since $(\lambda H-I)^{-1}$ can be computed - at least theoretically - by applying the results of Section 2.1, we therefore obtain a comprehensive description of $R'(\lambda;A)$.

THEOREM (2.D). Let $K_c: E \to E$ be a homeomorphism. Then the implications (2.10) are true, that is $\lambda \in \rho(A^{-1})$ if and only if $\lambda\hat{h}(\xi)-1 \neq o$ $(\xi \in \mathbb{R})$ and $w(\lambda H-I) = o$. Moreover, for any $\lambda \in \rho(A^{-1})$, the reduced resolvent $R'(\lambda;A)$ admits the representation

$$R'(\lambda;A)v = (\lambda H-I)^{-1}v + \lambda \frac{[K_o G(\lambda H-I)^{-1}v](o+)}{1-\lambda[K_o G(\lambda H-I)^{-1}g](o+)} (\lambda H-I)^{-1}g, \tag{2.12}$$

where $g(t) = e^{-kt}$ $(t \geq o)$ and $g(t) = o$ $(t < o)$.

<u>Proof.</u> (a) Necessity. If $o \neq \lambda \in \rho(A^{-1})$, then, by the implications (2.11), $\lambda A-I = \lambda(H-V)-I: E \to E$ is a homeomorphism. Thus, $\dim N(\lambda A-I) = o =$ codim $R(\lambda A-I)$, $N(A)$ and $R(A)$ denoting respectively the null space and

the range of an operator A. But

$$\text{ind } (\lambda A-I) = \dim N(\lambda A-I) - \text{codim } R(\lambda A-I) = \text{ind } (\lambda H-I-\lambda V),$$

and hence $\text{ind } (\lambda H-I-\lambda V) = o$. Since the operator λV is compact, it follows from Fredholm theory that $\text{ind } (\lambda H-I) = o$ too. We shall prove that $\lambda H-I$ is a Fredholm operator. Obviously, it is sufficient to show that $\dim N(\lambda H-I) < \infty$. Assume $N(\lambda H-I) \neq \{o\}$. Then there exists at least one element $o \neq u \in E$ such that $(\lambda H-I)u = o$, or equivalently that

$$(\lambda A-I)u(t) = - \lambda Vu(t) = - \lambda e^{-kt}(K_o Gu)(o+) \neq o.$$

Now, take any $v \in N(\lambda H-I)$, and let w denote the unique solution of the equation

$$(\lambda H-I)w(t) - \lambda Vw(t) = - \lambda e^{-kt} \qquad (t \geq o).$$

It is easily verified that $w := u/(K_o Gu)(o+)$ as well as $\tilde{w} := v/(K_o Gv)(o+)$ are solutions of this equation. Thus $v = \text{const}\cdot u$, since w is uniquely determined, and hence $\dim N(\lambda H-I) \leq 1$.

On the other hand, $\lambda H-I$ is a Fredholm operator if and only if $\lambda\hat{h}(\xi)-1 \neq o$ $(\xi \in \mathbb{R})$. Moreover, Fredholm operators $\lambda H-I$ generally satisfy the relation $\text{ind } (\lambda H-I) = w(\lambda H-I)$ (see M.G. Krein [34]). This completes the first part of the proof.

(b) Sufficiency. Let $\lambda \in \mathbb{C}$ be fixed such that $\lambda\hat{h}(\xi)-1 \neq o$ $(\xi \in \mathbb{R})$ and $w(\lambda H-I) = o$. Then, by an application of Theorem (2.A), the operator $\lambda H-I$ is a homeomorphism from E onto E. Given $v \in E$, we consider the equation $(\lambda A-I)w = v$, or equivalently $(\lambda H-I)w = v + \lambda Vw$ which can be rewritten as

$$w = (\lambda H-I)^{-1} [v + \lambda e^{-kt}(K_o Gw)(o+)]. \tag{2.13}$$

Apply K_oG on both sides, then after a rearrangement,

$$[1-\lambda(K_oG(\lambda H-I)^{-1} e^{-kt})(o+)](K_oGw)(o+) = (K_oG(\lambda H-I)^{-1}v)(o+). \tag{2.14}$$

If the brackets are not equal to zero, then $(K_oGw)(o+)$ is uniquely determined by (2.14), and hence (2.13) gives the representation (2.12). If, on the contrary, the brackets were equal to zero, then equation (2.14) would imply that $(K_oG(\lambda H-I)^{-1}v)(o+) = o$ for all $v \in E$. Choosing $v = \lambda e^{-kt}$ we would have established a contradiction. This concludes the proof.

To conclude this section we would like to give a simple application of the preceding theorem. We shall consider the special case where $k_o = o$, that is where $A = cG$.

COROLLARY (2.E). Let $A = cG$, where $o \neq c = |c| e^{i\tau} \in \mathbb{C}$ is an arbitrary constant. Then $\lambda \in \rho(A^{-1})$ if and only if $\operatorname{Re} \lambda e^{i\tau} < k/|c|$. In this case we have

$$R'(\lambda;A)v(t) = (\lambda H-I)^{-1}v(t) = -v(t) - c\lambda \int_o^t e^{-(k-c\lambda)(t-s)} v(s)ds.$$

<u>Proof.</u> According to Theorem (2.D) we have to verify the two conditions (i) $-1 + c\lambda/(k-i\xi) \neq o$ $(\xi \in \mathbb{R})$ and (ii) $w(\lambda H-I) = o$. Obviously, condi-

tion (i) holds for all λ satisfying $\mathrm{Re}\,\lambda e^{i\tau} \neq k/|c|$. Next we note that the winding number $w(\lambda H-I)$ is equal to the difference of the number of poles and the number of zeros of the meromorphic function

$$\lambda\hat{h}(z) - 1 = -[z + i(k-c\lambda)]/(z+ik)$$

within the half-plane $\mathrm{Im}\, z > o$. If $\mathrm{Re}\,\lambda e^{i\tau} < k/|c|$, then no poles and no zeros are contained in this half-plane which implies that $w(\lambda H-I) = o$. Finally, since the kernel $a(t) := c\lambda g(t)$ of the convolution operator A satisfies $a(t) = o$ $(t<o)$, we infer from the remark of Section 2.1 that

$$R'(\lambda;A)v(t) = -v(t) - \int_o^\infty h_+(t-s)v(s)ds \qquad (t \geq o).$$

Here $h_+(t)$ is the unique solution of equation (2.3) with $k(t)$ replaced by $-a(t)$. A simple computation yields $h_+(t) = c\lambda e^{-(k-c\lambda)t}$ $(t \geq o)$ which concludes the proof.

2.3. OPERATORS WITH RESOLVENT SETS CONTAINING A SECTOR

Notations are the same as in the preceding section. For the sake of simplicity we shall assume that $|c| > \|k_o\|_1$. This condition guarantees the existence of $K_c^{-1} \in B(E)$ because from the Neumann series it follows that

$$\|(cI+K_o)^{-1}\| = \|K_c^{-1}\| \leq (|c| - \|k_o\|_1)^{-1}.$$

Clearly, this condition implies also the more complicated conditions (2.8).

We shall be concerned in this section with the question whether the resolvent set $\rho(A^{-1})$ contains a symmetric sector S_θ of width 2θ and with vertex located in the origin of the complex plane,

$$S_\theta := \{\lambda \in \mathbb{C} | \lambda = |\lambda| e^{i\psi}, \quad o \le |\lambda| < \infty, \quad o \le |\psi| \le \theta\}.$$

This question is of significant interest with respect to the problem (P_ε), since (P_ε) leads to the consideration of an abstract differential equation of the type

$$\partial_x^2 u + \lambda u = f, \quad \lambda \in \mathbb{C}.$$

Solutions $u(x)$ of this equation are required to vanish as $x \to +\infty$. Since the general solution involves functions of the form $\exp(\pm \sqrt{-\lambda} x)$, we have to take care for a proper definition of $\sqrt{-\lambda}$. In order to make the square root of the complex number λ unique, we introduce a two-sheeted Riemannian surface, the upper sheet being connected with the lower sheet along the positive real axis. On the upper sheet we let $\lambda = |\lambda| e^{i\psi}$ $(o \le \psi < 2\pi)$, and hence

$$\sqrt{-\lambda} = i \cdot |\lambda|^{1/2} e^{i\psi/2} = -|\lambda|^{1/2} (\sin \psi/2 - i \cos \psi/2) \quad (o \le \psi < 2\pi). \quad (2.15)$$

Thus, if $o < \psi < 2\pi$, only the function $u(x) = e^{\sqrt{-\lambda} x}$ has the proper behaviour at $+\infty$.

In chapter 3 we shall be concerned with diverse Dunford-Taylor integrals over $u(x)$, where λ is running through an oriented path L surrounding the spectrum of the operator A^{-1} (see end of this section). Since, for definiteness, the cut-off line of the above Riemannian surface should not intersect with L, we have to require the semi-axis $\overline{\mathbb{R}}_+$ to be contained within the resolvent set $\rho(A^{-1})$. This is guaranteed by the condition $S_\theta \subset \rho(A^{-1})$.

We shall show that by an appropriate choice of the constant c it is always possible to find an angle $\theta \in (o,\pi)$ such that $S_\theta \subset \rho(A^{-1})$ holds.

To begin with, we consider the special case $A = cG$ explained in Corollary (2.E). If we let $c = |c|e^{i\tau}$ and $\lambda = |\lambda|e^{i\psi}$, then it follows from this corollary that

$$\lambda \in \rho(A^{-1}) \iff \cos(\psi+\tau) < k/|c||\lambda|. \tag{2.16}$$

Since $k>o$, this condition holds for any $\lambda\in\mathbb{C}$ satisfying $\pi/2 - \tau < \psi < 3\pi/2 - \tau$. Supposing that $\tau \in (\pi/2, 3\pi/2)$ and that $\theta_o := \min\{|\pi/2 - \tau|, 3\pi/2 - \tau\}$, then obviously $S_{\theta_o} \subset \rho(A^{-1})$.

The general case where $k_o \neq o$ is not as easy as the special case, because we have to replace (2.16) by the more complicated conditions (2.10). However, we shall prove an analogous result in Theorem (2.G). Our considerations in this direction are mainly governed by the following essential hypothesis.

(H1) SECTOR CONDITION. Let $|c| > \|k_o\|_1$ and define $q_o := \|k_o\|_1/|c|$. Let q by any real number such that $q_o < q < 1$. Then, for $c = |c|e^{i\tau}$, suppose that $\pi/2 + \arcsin q < \tau < 3\pi/2 - \arcsin q$, where $o < \arcsin q < \pi/2$. Now define $\theta \in (o, \pi/2)$ by

$$\theta := \min\{|\pi/2 - \tau + \arcsin q|,\ 3\pi/2 - \tau - \arcsin q\}.$$

This condition turns out to be sufficient in order that the sector S_θ is contained in $\rho(A^{-1})$. The proof is given in two steps. We shall again begin with the following

PROPOSITION (2.F). Let the Sector condition (H1) be satisfied. Then we have $\lambda\hat{h}(\xi)-1 \neq o$ $(\xi \in \mathbb{R})$ for all $\lambda \in S_\theta$.

<u>Proof.</u> Using (2.9) we see that the condition $\lambda\hat{h}(\xi)-1 \neq o$ $(\xi \in \mathbb{R})$ is equivalent with $|c||\lambda|\,[\cos(\psi+\tau) + i\sin(\psi+\tau)] + \mathrm{Re}[\lambda\hat{k}_o(\xi)] + i\,\mathrm{Im}[\lambda\hat{k}_o(\xi)] \neq k-i\xi$, and hence equivalent with

$$|k - |c||\lambda| \cdot \cos(\psi+\tau) - \mathrm{Re}[\lambda\hat{k}_o(\xi)]| > o \quad (\xi \in \mathbb{R}). \tag{2.17}$$

We have $\mathrm{Re}[\lambda\hat{k}_o(\xi)] \leq |\lambda|\,||k_o||_1$ because the Fourier transform $\hat{k}_o(\xi)$ is bounded by $||k_o||_1$ uniformly with respect to $\xi\in\mathbb{R}$. Assume now that $k - |c||\lambda| \cos(\psi+\tau) - |\lambda|\,||k_o||_1 > o$, then condition (2.17) holds. But $1 \geq -\cos(\psi+\tau) \geq q > q_o$ for all $\lambda \in S_\theta$, and hence

$$k - |c||\lambda| \cos(\psi+\tau) - |\lambda|\,||k_o||_1 \geq k + |c||\lambda|(q-q_o) > o. \tag{2.18}$$

This concludes the proof.

Now we are able to prove the main result of this section.

THEOREM (2.G). Let the Sector condition (H1) be satisfied. Then the resolvent set $\rho(A^{-1})$ includes the sector S_θ.

<u>Proof.</u> In view of Proposition (2.F) and of Theorem (2.D) it remains to show that $w(\lambda H-I) = o$ for all $\lambda \in S_\theta$. We first prove that $w(\lambda H-I)$ is a con-

stant on S_θ. To this purpose consider the curve $\zeta(\xi) \subset \mathbb{C}$ defined by

$$\zeta(\xi) := \lambda\hat{h}(\xi) - 1 = -1 + \lambda(c+\hat{k}_o(\xi))/(k-i\xi).$$

By definition, $w(\lambda H-I)$ is the winding number of $\zeta(\xi)$ with respect to the point $o\in\mathbb{C}$. Hence, using the Invariance-under-Homotopy theorem of the Brouwer index theory, we see that any curve $\zeta'(\xi)$ which is homotopic to $\zeta(\xi)$ in $\mathbb{C}\setminus\{o\}$ has the same winding number as $\zeta(\xi)$. Thus, let $\lambda,\lambda' \in S_\theta$, $\lambda \neq \lambda'$. Since the segment line $t\lambda' + (1-t)\lambda$ $(o \le t \le 1)$ is contained in S_θ, it follows by Proposition (2.F) that

$$h(t;\xi) := -1 + [t\lambda' + (1-t)\lambda](c+\hat{k}_o(\xi))/(k-i\xi) \neq o$$

for every $t \in [o,1]$. Therefore, $\zeta(\xi) = h(o;\xi)$ and $\zeta'(\xi) = h(1;\xi)$ are homotopic in $\mathbb{C}\setminus\{o\}$, and consequently $w(\lambda H-I) = w(\lambda' H-I)$.

To complete the proof we remark that for $\lambda=o$ we simply have $w(\lambda H-I) = o$. But $o \in S_\theta$.

REMARK. Proposition (2.F) and Theorem (2.G) remain true without any change, if in (H1) the condition on q, namely $q_o < q < 1$, is weakened to $q_o \le q < 1$. However, the stronger condition preserves the uniform boundedness of $\|R'(\lambda;A)\|$ with respect to $\lambda \in S_\theta$, whereas in case of $q_o = q$ the reduced resolvent need not to be bounded on the whole of S_θ as we will see in the next section.

We conclude this section with a simple result concerning the size of the resolvent set $\rho(A^{-1})$ in a neighbourhood of $o\in\mathbb{C}$.

LEMMA (2.H). Let B denote the open ball

$B := \{\lambda \,|\, |\lambda| < k(|c| + \|k_o\|_1)^{-1}\}$. Then $B \subset \rho(A^{-1})$.

Proof. The Neumann series applied to $(\lambda H-I)^{-1}$ yields

$$\|(\lambda H-I)^{-1}\| = \|-\sum_{\ell=o}^{\infty} (\lambda H)^{\ell}\| \leq (1 - |\lambda|\,\|H\|)^{-1}$$

for all $|\lambda| < \|H\|^{-1}$. But $\|H\| \leq \|h\|_1 \leq k^{-1}(|c| + \|k_o\|_1)$, and hence observing Theorem (2.D) and Theorem (2.A) the lemma is proved.

Summarizing to this point, we have shown that under the Sector condition (H1) the resolvent set $\rho(A^{-1})$ includes a path $L := \Gamma_\rho \cup \Gamma_+ \cup \Gamma_- \subset \rho(A^{-1})$, where

$$\Gamma_\rho := \{\lambda \in \mathbb{C} \,|\, \lambda = \rho e^{i\psi},\ \ \theta \leq \psi \leq 2\pi-\theta,\ \ o<\rho<k(|c| + \|k_o\|_1)^{-1}\},$$
$$\Gamma_\pm := \{\lambda \in \mathbb{C} \,|\, \lambda = re^{\pm i\theta},\ \ \rho \leq r < \infty\}.$$

An orientation of L has to be understood in the sense that L is traced from the lower half-plane to the upper half-plane in positive direction, see Fig. 1.

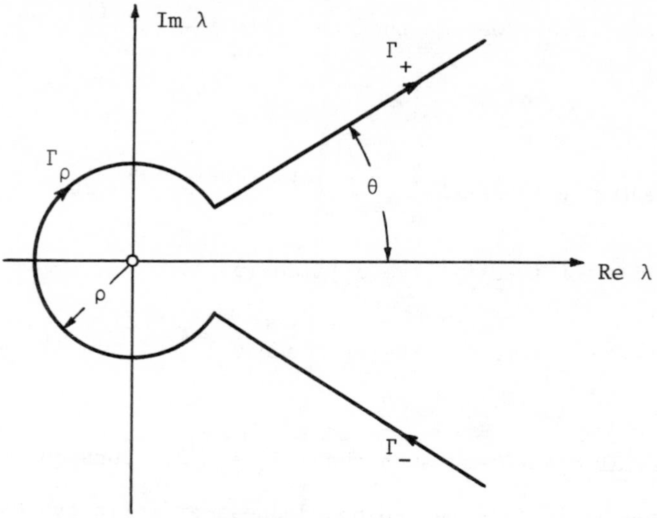

Fig.1

2.4. THE UNIFORM BOUNDEDNESS OF THE REDUCED RESOLVENT

Notations in this section are the same as before. We have seen that the Sector condition (H1) guarantees the existence of an unbounded component of the resolvent set $\rho(A^{-1})$ which includes the sector S_θ. Now we are interested in establishing sufficient conditions on A which guarantee that the reduced resolvent $R'(\lambda;A)$ is bounded uniformly with respect to λ on the whole sector S_θ.

We shall show that the Sector condition (H1) is exactly the proper condition. The proof involves differentiability properties of the norm $||\cdot||$ of the Banach space E, and therefore it has familiarities with the theory of duality mappings and of accretive operators. We shall come back to this topic in Section 2.6.

First we shall consider again the special case of $A = cG$. For

$c = |c|e^{i\tau}$ we let $\pi/2 < \tau < 3\pi/2$, and $\theta_o = \min\{|\pi/2 - \tau|, 3\pi/2 - \tau\}$. Using Corollary (2.E), we see that for all $\lambda = |\lambda|e^{i\psi} \in S_{\theta_o}$ the following estimates hold.

$$\begin{aligned} ||R'(\lambda;A)|| &\leq 1 + |c||\lambda| \int_o^\infty e^{-(k-|c||\lambda|\cos(\psi+\tau))t}\,dt \\ &\leq 1 + |c||\lambda| \ (k-|c||\lambda|\cos(\psi+\tau))^{-1} \\ &\leq 1 + |c||\lambda| \ (k+|c||\lambda|\cos\theta_o)^{-1}. \end{aligned}$$

Take the special case $\tau=\pi$ such that $\theta_o = \pi/2$. Consequently, if $\psi = \pm\theta_o$, then the above estimate gives no boundedness at infinity. However, if θ is any number less than θ_o, then it follows that the reduced resolvent is bounded uniformly on the sector S_θ by $||R'(\lambda;A)|| \leq 1 + (\cos\theta)^{-1}$.

The main result of this section is

THEOREM (2.I). Let the Sector condition (H1) be satisfied. Then, if q_o and q are the constants fixed in (H1), the reduced resolvent $R'(\lambda;A)$ is uniformly bounded with respect to $\lambda \in S_\theta$ by

$$||R'(\lambda;A)|| \leq 1 + ||K_c|| \,/\, |c|(q-q_o).$$

To prove this theorem we proceed in several steps. First of all we shall prove a well-known result concerning the differentiability of the norm $||\cdot||$.

PROPOSITION (2.K). Let $\Delta(v,f;r)$ denote the difference quotient of $||v||$ in the direction f, $\Delta(v,f;r) := (||v+rf|| - ||v||)/r$ $(r \neq o)$. Then the mapping $r \to \Delta(v,f;r)$ is monotonic increasing for all $r>o$.

Proof. Let $r, s>o$, and assume that $r>s$. Then there exists a number $t \in (o,1)$ such that $s = tr$. Since the function $t \to ||tv + f||$ is convex we obtain

$$\begin{aligned}
&\Delta(v,f;r) - \Delta(v,f;s) \\
&= (rs)^{-1} \, (||sv + srf|| - ||rv + rsf|| + (r-s)\,||v||) \\
&= (tr^2)^{-1} \, (t\,||rv + r^2 f|| - ||[t+(1-t)]\, rv + tr^2 f|| + r(1-t)\,||v||) \geq o,
\end{aligned}$$

as claimed.

The difference quotient $\Delta(v,f;r)$ is bounded from below by $-||f||$. This, together with the monotonicity just proved, shows that the right derivative

$$D_+(v,f) := \lim_{r\to o+} \Delta(v,f;r)$$

exists. Using this result, we are in the position to prove the following lemma.

LEMMA (2.L). Let $c = |c|e^{i\tau}$ and $\lambda = |\lambda|e^{i\psi}$ be any given numbers such that

$$k - |c|\,|\lambda| \cos(\psi+\tau) > o. \tag{2.19}$$

Let $|c| > ||k_0||_1$, then we have for any $v \in E_0^1$

$$(k - |c|\,|\lambda| \cos(\psi+\tau) - |\lambda|\,||k_0||_1)||v|| \leq ||K_c||\,||\lambda v - A^{-1}v||. \tag{2.20}$$

Proof. Let $v \in E_o^1$ be fixed, and define $w := \lambda v - A^{-1}v = \lambda v - K_c^{-1}G^{-1}v \in E$. Thus $K_c w = \lambda K_c v - G^{-1}v$ because $K_c : E \to E$ is a homeomorphism. Let $r>o$, then we obtain $-r^{-1}(v-r(\lambda K_c v - G^{-1}v) - v) = K_c w$. Defining $f := -|c||\lambda|e^{i(\psi+\tau)}v+G^{-1}v$ we infer that

$$\Delta(v,f;r) - |\lambda|\|k_o\|_1 \|v\| \leq \|K_c\| \|\lambda v - A^{-1}v\|. \tag{2.21}$$

Next, we shall prove a lower bound for $D_+(v,f)$. Let $\beta := k + r^{-1} - |c||\lambda|e^{i(\psi+\tau)}$, and let $\tilde{v} := (\beta+\partial_t)v \in E$. Since, for any $r>o$, $\operatorname{Re}\beta \geq k - |c||\lambda| \cos(\psi+\tau) > o$ we will find that

$$v(t) = \int_o^t e^{-\beta(t-s)}\tilde{v}(s)ds,$$

and hence $\|v\| \leq \|\tilde{v}\| / \operatorname{Re}\beta$. Using this estimate together with $\|\tilde{v}\| \geq |\beta| \|v\| - \|\partial_t v\|$ we will see that

$$\Delta(v,f;r) \geq \frac{(k-|c||\lambda| \cos(\psi+\tau))}{r\cdot\operatorname{Re}\beta} (r|\beta|\|v\| - r\|\partial_t v\|).$$

Therefore $D_+(v,f) \geq \|v\|(k-|c||\lambda| \cos(\psi+\tau))$, since $\lim\limits_{r\to o+} r\cdot\operatorname{Re}\beta = 1 = \lim\limits_{r\to o+} r\cdot|\beta|$. Combining this estimate with (2.21), we obtain the desired result.

We now assume the Sector condition (H1) to be satisfied which implies that inequality (2.18) is valid. In particular, this shows that $k - |c||\lambda| \cos(\psi+\tau) > |\lambda|\|k_o\|_1 \geq o$, and hence condition (2.19) holds. A combination of the inequalities (2.18) and (2.20) yields $o<k + |c||\lambda| (q-q_o) \leq \|\lambda I-A^{-1}\|$. Going back to Theorem (2.G) we thus have

proved:

COROLLARY (2.M). Let the Sector condition (H1) be satisfied. Let q_o, q and θ be the constants specified in (H1). Then on each straight line $\lambda = |\lambda| e^{i\psi}$ $(o \le |\psi| \le \theta)$ the function $||R(\lambda;A^{-1})||$ is bounded from above by a monotonic decreasing function of $|\lambda|$:

$$||R(\lambda;A^{-1})|| \le ||K_c||/[k + |c||\lambda| (q-q_o)].$$

<u>Proof of Theorem</u> (2.I). If $\lambda=o$, then $R'(\lambda;A) = I$ and there is nothing to be proved. If $o \neq \lambda \in S_\theta$, an application of Proposition (2.B) yields

$$R(\lambda;A^{-1}) = \lambda^{-1}(\lambda A-I)^{-1} (\lambda A-I+I) = \lambda^{-1}I + \lambda^{-1}R'(\lambda;A). \qquad (2.22)$$

This implies using Corollary (2.M) that $||R'(\lambda;A)|| \le 1 + |\lambda| \, ||R(\lambda;A^{-1})|| \le 1 + ||K_c|| / |c| (q-q_o)$ as claimed.

We conclude this section by a result concerning the boundedness of $R'(\lambda;A)$ along the path L introduced in Section 2.3.

COROLLARY (2.N). Let the Sector condition (H1) be satisfied. Let $\Omega \subset \mathbb{C}$ denote the open domain surrounded by L and including the origin. Then there exists a constant C independent of λ such that $||R'(\lambda;A)|| \le C$ for every $\lambda \in \overline{\Omega}$.

<u>Proof.</u> This can be seen immediately by Theorem (2.I) and by an application of the Neumann series

$$(\lambda A-I)^{-1} = -\sum_{\ell=0}^{\infty} \lambda^{\ell} A^{\ell} \tag{2.23}$$

in the ball $o \leq |\lambda| \leq \rho < k/(|c| + ||k_o||_1)$.

2.5. ESTIMATES FOR $R'(\lambda;A)e^{-kt}$

This section may be motivated by the following observation. Consider the identity

$$R'(\lambda;A)v = \lambda^{-1}R'(\lambda;A)(\lambda A-I+I)A^{-1}v = \lambda^{-1}A^{-1}v + \lambda^{-1}R'(\lambda;A)A^{-1}v, \tag{2.24}$$

which is valid for all $o \neq \lambda \in \rho(A^{-1})$ and for all $v \in E_o^1$. Inasmuch as $\lambda \in S_\theta$, the reduced resolvent is uniformly bounded, and hence from (2.24) it easily follows that

$$||R'(\lambda;A)v|| = O(|\lambda|^{-1}) \quad \text{as} \quad |\lambda| \to \infty \quad (\lambda \in S_\theta), \tag{2.25}$$

provided that $v \in E_o^1$. However, in applications it frequently turns out that $v \in E^1$ (which is the subset of E of absolutely continuous functions such that $\partial_t v \in E$) but $v(o) \neq o$. Therefore, (2.25) does not apply directly to $v \in E^1$. However, let $\tilde{v}(t) := v(t) - v(o)e^{-kt}$. Now inequality (2.25) is valid for $\tilde{v}$ because of $\tilde{v} \in E_o^1$. Clearly, this simple trick fails to be helpful, if we cannot find an estimate for the remainder $R'(\lambda;A)e^{-kt}$ which is better than the one obtained from Theorem (2.I).

In this section we shall prove such an appropriate estimate. Contrary to former results on boundedness this estimate will depend upon the specific choice of the spaces $E \in \mathcal{E}$. The best result (in view of the growth rate at infinity) will be obtained in the space L_+^1. In the spaces $E \subseteq L_+^\infty$ an improvement of the results known from Theorem (2.I) is given by pointwise estimates (with respect to $t \in \overline{\mathbb{R}}_+$).

We shall begin with a reduction of $R(\lambda;A^{-1})e^{-kt}$ to a <u>Volterra integral</u> equation.

PROPOSITION (2.P). Let the Sector condition (H1) be satisfied. Then we have

$$R(\lambda;A^{-1})e^{-kt} = \lambda^{-1}e^{-kt} - v_o(t) \in E_o^1 , \tag{2.26}$$

where v_o is the unique solution of the Volterra integral equation

$$v_o(t) = \lambda^{-1}e^{-(k-c\lambda)t} + \lambda \int_o^t e^{-(k-c\lambda)(t-s)} (K_o v_o)(s)ds. \tag{2.27}$$

<u>Proof.</u> First we shall prove that equation (2.27) has a unique solution $v_o \in E$. Consider the operator $E \ni v \mapsto Lv \in E$ where

$$Lv := \lambda \int_o^t e^{-(k-c\lambda)(t-s)} (K_o v)(s)ds.$$

Using (2.18), we get $||Lv|| \leq |\lambda| \, ||k_o||_1 \, (k - |c||\lambda| \cos(\psi+\tau))^{-1} < 1$, and hence we may apply the fixed-point contraction theorem in order to conclude that there exists a unique $v_o \in E$ solving equation (2.27). In particular, $v_o \in E^1$ since L maps the space E into E_o^1.

Next, an elementary calculation shows that v_o solves the initial-value problem

$$\left.\begin{aligned} &\partial_t v_o + (k-c\lambda)v_o - \lambda K_o v_o = o \quad (t>o), \\ &v_o(o) = \lambda^{-1}. \end{aligned}\right\}$$

Letting $v(t) := \lambda^{-1}e^{-kt} - v_o(t) \in E_o^1$ one obtains from the first equation $K_c e^{-kt} = \lambda K_c v - G^{-1}v_o$. Multiplying from the left by K_c^{-1} and observing that $\lambda \in \rho(A^{-1})$, we finally get $R(\lambda;A^{-1})e^{-kt} = v(t) = \lambda^{-1}e^{-kt} - v_o(t)$. Hence the proposition is proved.

The following theorem establishes bounds for $R'(\lambda;A)e^{-kt}$ in the spaces L_+^p $(1 \le p < \infty)$.

THEOREM (2.Q). Let the Sector condition (H1) be satisfied. Let $E := L_+^p$ $(1 \le p < \infty)$. Then

$$\left.\begin{aligned} ||R'(\lambda;A)e^{-kt}||_p &\le p^{-1/p}(k+|c||\lambda|)^{(p-1)/p}\,(k+|c||\lambda|(q-q_o))^{-1} \\ &= O(|\lambda|^{-1/p}) \end{aligned}\right\} \tag{2.28}$$

as $|\lambda| \to \infty$ $(\lambda \in S_\theta)$.

<u>Proof.</u> According to Proposition (2.P) we have for any $E \in \mathcal{E}$

$$||v_o|| \le |\lambda|^{-1}||e^{-(k-c\lambda)t}|| + ||v_o|||\lambda|||k_o||_1\,(k-|c||\lambda|\cdot\cos(\psi+\tau))^{-1},$$

and hence

$$||v_o|| \le |\lambda|^{-1}\,\frac{k-|c||\lambda|\cos(\psi+\tau)}{k-|c||\lambda|\cos(\psi+\tau)-|\lambda|||k_o||_1}\,||e^{-(k-c\lambda)t}||. \tag{2.29}$$

Next, combination equation (2.22) with (2.26) we will see that

$$R'(\lambda;A)e^{-kt} = -\lambda v_o(t). \tag{2.30}$$

Making use of the estimate $\|e^{-(k-c\lambda)t}\|_p \le [p(k-|c||\lambda| \cos(\psi+\tau))]^{-1/p}$ the required result is immediately obtained from (2.29).

REMARK. Observe that $\lim_{p\to\infty} p^{-1/p} = 1$. After passing in relation (2.28) to the limit $p \to +\infty$, we see that Theorem (2.Q) is also valid in the spaces $E \subseteq L_+^\infty$, since $\|e^{-(k-c\lambda)t}\|_\infty = 1$. The resulting estimate is only $O(1)$ for $|\lambda| \to \infty$ and hence not better than the one proved in Theorem (2.I).

We now proceed to the derivation of a pointwise estimate. Because of the relation (2.30) the starting point of the process will be the Volterra integral equation (2.27). Since there exists a unique solution v_o, we substitute $v_o(t) = e^{-(k-c\lambda)t}v(t)$ in (2.27) in order to get rid of the exponential function. This yields

$$v(t) = \lambda^{-1} + \lambda \int_0^t \int_0^\infty k_o(s-r)e^{(k-c\lambda)(s-r)}v(r)drds.$$

Passing to the moduli of the functions we will find that

$$|v(t)| \le |\lambda|^{-1} + K|v(t)| :=$$

$$|\lambda|^{-1} + |\lambda| \int_0^t \int_{-\infty}^s |k_o(r)|\ e^{\mathrm{Re}(k-c\lambda)r}|v(s-r)|\ drds.$$

We shall apply the method of superfunctions in order to obtain an a priori estimate of $|v(t)|$. To this purpose we will prove:

PROPOSITION (2.R). Let the Sector condition (H1) be satisfied. Let

$o \neq \lambda \in S_\theta$, and suppose that $k_o(t) = o$ for all $t>o$. Let $w(t) := |\lambda|^{-1} e^{|\lambda| \|k_o\|_1 t}$ $(t \geq o)$, then the following inequality holds

$$|\lambda|^{-1} + |\lambda| \int_o^t \int_{-\infty}^s |k_o(r)| \, e^{\mathrm{Re}(k-c\lambda)r} \, w(s-r)\,dr\,ds \leq w(t) \qquad (t \geq o).$$

<u>Proof.</u> Note that for all $\lambda \in S_\theta$ we have $\beta := \mathrm{Re}(k-c\lambda) - |\lambda| \|k_o\|_1 > o$ (see (2.18)). Using integration by parts, we readily get

$$\begin{aligned} |\lambda|^{-1} + Kw(t) &= |\lambda|^{-1} + (|\lambda| \|k_o\|_1)^{-1} \, (e^{|\lambda| \|k_o\|_1 t} - 1) \int_{-\infty}^{o} |k_o(r)| \, e^{\beta r} \, dr \\ &\leq |\lambda|^{-1} \, e^{|\lambda| \|k_o\|_1 t} = w(t). \end{aligned}$$

The proof is complete.

Besides the just proved inequality $|\lambda|^{-1} + Kw(t) \leq w(t)$ we have already seen that $|v(t)| \leq |\lambda|^{-1} + K|v(t)|$. Therefore, $|v(t)| \leq w(t)$ $(t \geq o)$ by an application of a well-known principle on integral inequalities (see W. Walter [49]). Hence we have proved the following theorem.

THEOREM (2.S). Let the Sector condition (H1) be satisfied. Let $k_o(t) = o$ for all $t>o$. Then we have for all $\lambda = |\lambda| e^{i\psi} \in S_\theta$ and all $t \geq o$:

$$|R'(\lambda;A)e^{-kt}| \leq e^{-(k-|c||\lambda|\cos(\psi+\tau) - |\lambda| \|k_o\|_1)t} \leq e^{-(k+|c||\lambda| \, (q-q_o))t}.$$

REMARK. Problems with a kernel function k_o satisfying $k_o(t) = o$ $(t>o)$ are of physical interest in the theory of heat conduction in materials with memory (see H. Grabmüller [22]).

2.6 MONOTONICITY METHODS

In this section E will be a complex normed space with norm $\|\cdot\|$, and E^* will denote its topological dual. As a well-known fact, derivatives of the norm are closely related to the duality mapping $F: E \to 2^{E^*}$ (which is in general multi-valued) defined by

$$F(v) := \{\ell \in E^* \mid \ell(v) = \|v\|^2 = \|\ell\|^2\} \qquad (v \in E).$$

Subsequently we shall reconsider the proof of Lemma (2.L) using now some aspects arising from properties of the duality mapping F. We shall start with a brief report upon well-known facts of F.

First of all, by an application of the Hahn-Banach theorem it is clear that $F(v) \neq \emptyset$ for every $v \in E$.

As before, let $\Delta(v,f;r) = r^{-1}(\|v + rf\| - \|v\|)$ denote the difference quotient of $\|v\|$ in direction f. The right derivative $D_+(v,f) := \lim_{r \to 0+} \Delta(v,f;r)$ exists which was proved in Section 2.4. A similar argumentation shows that the left derivative $D_-(v,f) := \lim_{r \to 0-} \Delta(v,f;r)$ also exists. If the norm is Gâteaux-differentiable at $\|v\|$ then the common value $D_-(v,f) = D_+(v,f)$ is the Gâteaux derivative. In general, $D_-(v,f) \neq D_+(v,f)$. However, we will prove:

PROPOSITION (2.T). Let $v \in E$, and let $\ell \in F(v)$. Then we have

$$D_-(v,f)\|v\| \leq \operatorname{Re} \ell(f) \leq D_+(v,f)\|v\| \quad \text{for every } f \in E. \tag{2.31}$$

Proof. Let $r \neq 0$. Then by the relations $\ell(v) = \|v\|^2 = \|\ell\|^2$ it follows that

$$\|v\|^2 + r \operatorname{Re} \ell(f) = \operatorname{Re} \ell(v+rf) \leq \|v\| \|v+rf\|.$$

From this we obtain $\operatorname{Re} \ell(f) \leq \Delta(v,f;r)\|v\|$ $(r>o)$ and $\Delta(v,f;r)\|v\| \leq \operatorname{Re} \ell(f)$ $(r<o)$. Passing to the limits $r \to o+$ respectively $r \to o-$, the assertion (2.31) is proved.

A normed space E is called <u>strictly</u> <u>convex</u>, if the unit ball $S \subset E$ is strictly convex that is, if $u,v \in \partial S$ and $t \in (o,1)$ imply that $tu + (1-t)v \in \operatorname{int} S$.

In any normed space E having a strictly convex dual E^*, the duality mapping $F: E \to E^*$ is single-valued (we refer to M.M. Vainberg [45] for instance). The most famous examples of such spaces are the spaces $L^p(\Omega)$ $(1<p<\infty)$ with $\Omega \subset \mathbb{R}^n$ $(n \geq 1)$ being an arbitrary (finite or infinite) measurable set. In these spaces the duality mapping $F: L^p(\Omega) \to L^q(\Omega)$ $(p^{-1} + q^{-1} = 1)$ is given by

$$F(v)(f) = \|v\|_p^{2-p} \int_\Omega |v(x)|^{p-2} \overline{v(x)} f(x)dx. \tag{2.32}$$

Moreover, since the $L^p(\Omega)$ norm $(1<p<\infty)$ is Fréchet-differentiable it follows from (2.31) that

$$D_-(v,f) \|v\|_p = \operatorname{Re} F(v)(f) = D_+(v,f) \|v\|_p. \tag{2.33}$$

for all $v,f \in L^p(\Omega)$.

If in particular $\Omega = \overline{\mathbb{R}}_+$, then $L^p(\Omega) = L^p_+$, and hence the above statements get an importance in view of the analysis developed in the previous sections. For the sake of simplicity we shall focus our interest exceptionally to these spaces. Hence, for the present we shall assume that $E = L_+^p$ $(1<p<\infty)$.

An operator $A: E \supset D(A) \to E$ (linear or nonlinear) is called accretive (or F-monotone) if and only if $\operatorname{Re} F(v)(Av) \geq o$ for all $v \in D(A)$. It is called strongly accretive, if there is a continuous monotonic increasing function $\phi: \overline{\mathbb{R}}_+ \to \overline{\mathbb{R}}_+$ satisfying $\phi(o) = o$ such that $\operatorname{Re} F(v)(Av) \geq \phi(\|v\|)$ for all $v \in D(A)$.

LEMMA (2.U). The operator $G^{-1} = kI + \partial_t: E_o^1 \to E$ is strongly accretive in the spaces $E = L_+^p$ $(1<p<\infty)$. More precisely, $\operatorname{Re} F(v)(G^{-1}v) \geq k\|v\|_p^2$ for all $v \in E_o^1$.

Proof. Let $v \in E_o^1$ be fixed. We have
$\partial_t |v(t)|^p = p|v(t)|^{p-2} \cdot \operatorname{Re}(\overline{v(t)} \cdot \partial_t v(t))$.
Therefore, using (2.32) it follows that

$$\begin{aligned}\operatorname{Re} F(v)(G^{-1}v) &= \|v\|_p^{2-p} \int_o^\infty [k|v(t)|^p + |v(t)|^{p-2} \cdot \operatorname{Re}(\overline{v}\cdot\partial_t v)(t)]dt \\ &= k\|v\|_p^2 + p^{-1}\|v\|_p^{2-p} \int_o^\infty \partial_t |v(t)|^p dt \\ &\geq k\|v\|_p^2 ,\end{aligned}$$

since $v(o) = o$. The Lemma is proved.

REMARK. In the Hilbert space $E = L_+^2$ we also have

$$\begin{aligned}\operatorname{Re} F(w)(Gw) &= \int_o^\infty \operatorname{Re}(\overline{w}Gw)(t)dt = \operatorname{Re} F(Gw)(w) \\ &= \operatorname{Re} F(v)(G^{-1}v) \geq k\|v\|_2^2 = k\|Gw\|_2^2 \geq o\end{aligned}$$

since $G: E \to E_o^1$ is a bijection. Consequently, G is accretive in L_+^2. This result fails to be true in the spaces L_+^p, $p \neq 2$.

Turning back to the resolvent $R(\lambda;A^{-1})$ we shall prove another version of Lemma (2.L) under slightly modified assumptions.

LEMMA (2.V). Let $E = L_+^p$ $(1<p<\infty)$. Let $c = |c|e^{i\tau}$ and $\lambda = |\lambda|e^{i\psi}$, and suppose that $|c| > ||k_o||_1$. Then inequality (2.20) holds.

Proof. In fact, from the equation $-r^{-1}[v-r(\lambda K_c v-G^{-1}v)-v] = K_c(\lambda v-A^{-1}v)$ (see proof of Lemma (2.L)) we infer that

$$r^{-1}(||v-r(\lambda K_c v-G^{-1}v)||_p - ||v||_p) \leq ||K_c|| \, ||\lambda v-A^{-1}v||_p.$$

Together with Lemma (2.U) and with $\operatorname{Re} F(v)(-\lambda K_c v) \geq -(|c||\lambda| \cos(\psi+\tau) + |\lambda| \, ||k_o||_1)||v||_p^2$, this inequality implies the desired estimate (2.20).

REMARK. The same proof can be performed in the space $E = L_+^1$. Note that the (multivalued) duality mapping F of L_+^1 to its dual L_+^∞ is given by

$$F(v)(f) = ||v||_1 \int_o^\infty \operatorname{sign}[v(t)]f(t)dt$$

where we assume that E is a real space, and where

$$\operatorname{sign}[v] = \begin{cases} 1 & \text{for } v > o, \\ [-1,1] & \text{for } v = o, \\ -1 & \text{for } v < o. \end{cases}$$

The proof of Lemma (2.V) admits obvious generalizations to arbitrary (linear or nonlinear) strongly accretive operators G^{-1}. In view of this fact we are motivated to give the preference to the monotonicity method over the approach used in Section 2.4.

BIBLIOGRAPHICAL NOTES

Sec. 2.1. The study of convolution operators K has its beginning in the first twenty years of this century. It appeared at first that the discovery of explicit analytic formulas for the solutions of the equation $(I+K)v = f$ in general was impossible. However, this point of view was refuted by the approach of N. Wiener and E. Hopf [50] who succeeded in developing the theory and technique for analytic solutions by using the idea of factorization. I.M. Rapoport [42] was the first who made clear that the winding number plays an essential role in the theory of convolution operators. A sufficient general theory - in view of the applicability to our singular perturbation problem - was then achieved by M.G. Krein [34]. He introduced the collection of function spaces E in the analysis of convolution operators. Theorem (2.A) and the representation formulae (2.3), (2.4), (2.5) and (2.6) are due to him.

Sec. 2.2. The study of the resolvent set $\rho(A^{-1})$ is closely connected with the theory of integro-differential equations involving convolution operators. This theory has been studied by many authors. First hints to an appropriate approach can already be found in the work of M.G. Krein [34]. Nearly at the same time, A. Ahiezer, N. Ahiezer and G. Lyubarskii [1] succeeded in solving second order integro-differential equations by using the Wiener-Hopf technique. Later, E. Gerlach [20] has treated the general case of n-th order equations by a consequent application of Krein's results.

Lemma (2.C) and Theorem (2.D), and also the introduction of the operator G, are essentially due to him. However, the proof of Theorem (2.D), and especially the representation formula (2.12) are new in this context.

Sec. 2.3 and 2.4. According to T. Kato [31], a closed linear operator A is called <u>sectorial</u>, if its numerical range is a subset of a conic sector $|\arg(\lambda-\gamma)| \leq \theta < \pi/2$ $(\lambda\in\mathbb{C})$ with vertex $\gamma\in\mathbb{C}$, say S_θ'. Clearly, in this case the operator $\gamma I + A$ is accretive. Moreover, if $\gamma I + A$ is maximal accretive, then it can be proved that the sector S_θ' contains the spectrum of A; in other words the resolvent set covers the exterior of S_θ'. By reflecting S_θ' at the imaginary axis of the complex plane, one obtains the position of the spectrum of A in case that $\gamma I + A$ is a maximal dissipative operator. This is just the situation which is required by the sector property of the operator $K_c^{-1} G^{-1}$. The sector property was conjectured by the author [23], where no conditions for its fulfilment were given. The Sector condition (H1) and its consequences for the uniform boundedness of the reduced resolvent seem to be new. Only Proposition (2.K) has been adapted from V. Barbu [3].

Sec. 2.5. The estimate (2.28) was established by the author [23] in the special case of $p = 2$.

Sec. 2.6. Duality mappings were first considered by M.M. Vainberg [46]. A detailed account, especially for multivalued duality mappings, was later given by F.E. Browder [5,6,7] and others. Today, a nearly complete representation of the theory can be found in the monographs of M.M. Vainberg [45] and V. Barbu [3]. From the latter we have adapted the proof of Proposition (2.T). These monographs also contain a good part of the theory of accretive operators which have been introduced by K.O. Friedrichs [19] and R.S. Phillips [39, 40].

3 Abstract two-point boundary value problems in Banach spaces

In this chapter we shall consider certain aspects of an operational calculus based on the results which were found in chapter 2 for the resolvent of convolution operators. In particular, we shall discuss the analysis of different analytic functions depending upon the closed operator A^{-1} introduced before. This discussion gives us an opportunity to develop that part of the machinery which will be needed for the treatment of two important classes of abstract two-point boundary value problems arising from singular perturbation calculus.

The first section contains further notations of abstract function spaces, and the fundamental Root definition (H2) which will guarantee a uniform interpretation of complex square roots frequently appearing in the subsequent analysis. Furthermore, the operator family $S_o(y)$ will be introduced by means of a Dunford-Taylor integral. This family has a universal significance because it is the core of the whole existence theory for the singular perturbation problem.

In the second section we shall formulate the two-point boundary value problems which are of our peculiar interest. In connection with these problems we introduce another two operator families which turn out to be helpful for the construction of solutions to inhomogeneous problems.

Section 3.3 brings a brief report upon semigroups of continuous linear operators and their relationship to the operator family $S_o(y)$ which is associated with an infinitesimal generator.

The last two sections are concerned with the existence theory for the two-point boundary value problems under consideration. General existence theorems involving right-hand side functions of bounded variation will be established in Section 3.4 whereas in Section 3.5 we will show that the smoothing effect of Hölder continuous right-hand side functions leads to a considerable improvement of the existence results.

3.1 DEFINITION AND FUNDAMENTAL PROPERTIES OF THE OPERATOR FAMILY $S_o(y)$

In this and the following sections E will denote a fixed Banach space of the collection $\mathcal{E}$ introduced in Section 2.1. We shall consider several operator-valued functions which are defined on the semi-axis $\mathbb{R}_+$ (respectively on $\overline{\mathbb{R}}_+$) and which have values in the Banach algebra $B(E)$.

In particular, we will denote by $C(\mathbb{R}_+;B(E))$ the class of all those functions which are strongly continuous. By $C^k(\mathbb{R}_+;B(E))$ $(k\in\mathbb{N})$ we will denote the class of functions whose strong derivatives of order less or equal than k are elements of $C(\mathbb{R}_+;B(E))$. Furthermore, we will agree with the usual notation

$$C^\infty(\mathbb{R}_+;B(E)) := \bigcap_{k\in\mathbb{N}} C^k(\mathbb{R}_+;B(E)).$$

The symbol $C_o(\mathbb{R}_+;B(E))$ will denote the subclass of functions $f \in C(\mathbb{R}_+;B(E))$ having a limit $\lim_{y\to+\infty} ||f(y)|| = o$.

Furthermore, $L^p(\mathbb{R}_+;B(E))$ $(1 \le p < \infty)$ will denote the Banach space of strongly measurable and Bochner integrable functions which are defined a.e. on $\mathbb{R}_+$ and which assume values in $B(E)$ having a finite norm

$$||f||_{p,B(E)} := (\int_o^\infty ||f(y)||^p_{B(E)}\, dy)^{1/p} ,$$

and finally, $L^\infty(\mathbb{R}_+;B(E))$ will denote the Banach space of strongly measurable and essentially bounded functions defined a.e. on $\mathbb{R}_+$ such that

$$||f||_{\infty,B(E)} := \text{ess.} \sup_{y>0} ||f(y)||_{B(E)} < \infty.$$

If there is no danger of confusion we shall omit the index $B(E)$ from the operator norm $||\cdot||_{B(E)}$.

The same notations as in chapter 2 will be used for the operators G, $K_c = cI + K_o$, and $A = GK_c$.

This section is mainly devoted to the study of an operator family which plays an important role in the existence theory for the singular perturbation problem. Let $L\subset\mathbb{C}$ be the oriented path described in Fig. 1 of Section 2.3. Then let us define the operator family $S_o(y)$ $(y>0)$ by the Dunford-Taylor integral

$$S_o(y) := (2\pi i)^{-1} \int_L e^{y\sqrt{-\lambda}} R'(\lambda;A)\lambda^{-1} d\lambda, \qquad (3.1)$$

where $R'(\lambda;A) = (\lambda A-I)^{-1}$ is the reduced resolvent of the operator $A^{-1} = K_c^{-1}G^{-1}$.

Suppose that the Sector condition (H1) holds. Then it is immediately obvious that the above definition makes sense, if the value of the square root $\sqrt{-\lambda}$ is fixed in a proper way. This will be done by using the definition of $\sqrt{-\lambda}$ given in (2.15), to which we shall refer in the sequel as to the

(H2) ROOT DEFINITION. Let $\lambda = |\lambda|e^{i\psi}$ be any complex number. Then define the value of the square root $\sqrt{-\lambda}$ by $\sqrt{-\lambda} = -|\lambda|^{1/2}(\sin\psi/2 - i\cos\psi/2)$

$(o \leq \psi < 2\pi)$.

Clearly, the definition (3.1) of $S_o(y)$ does not depend upon the particular choice of the contour L concerning the size of ρ within its limits o and $k(|c| + ||k_o||_1)^{-1}$, and the size of the angle within zero and θ. This may be justified by an application of Cauchy's theorem.

The importance of $S_o(y)$ is based on its relationship to the following homogeneous two-point boundary value problem

$$\left.\begin{aligned} &A\partial_y^2 u(y) + u(y) = o \quad (o<y<\infty), \\ &u(o) = u_o \quad \text{and} \quad \lim_{y\to+\infty} ||u(y)|| = o. \end{aligned}\right\} \tag{3.2}$$

Indeed, the following main theorem of this section contains an existence result for the problem (3.2)

THEOREM (3.A). Let the Sector condition (H1) be satisfied, and suppose that the Root definition (H2) holds. Then we have

(i) $S_o(\cdot) \in C^\infty(\mathbb{R}_+;B(E)) \cap L^p(\mathbb{R}_+;B(E)) \cap C_o(\mathbb{R}_+;B(E)) \quad (1 \leq p < \infty)$,

(ii) $A\partial_y^2 S_o(y) + S_o(y) = o \quad$ for all $\quad y > o$,

(iii) $\lim_{y\to o+} ||S_o(y)v - v|| = o \quad$ for every $\quad v \in E_o^1$.

We shall divide the proof of this theorem into several steps. Let us first perform some auxiliary considerations which will repeatedly appear in our subsequent calculations.

Using the Root definition (H2) for numbers λ lying on the contour $L = \Gamma_\rho \cup \Gamma_+ \cup \Gamma_-$, we will immediately see that

$$\sqrt{-\lambda} = \left\{ \begin{array}{ll} -\rho^{1/2}(\sin\psi/2 - i\cos\psi/2), \text{ if } \lambda = \rho e^{i\psi} \in \Gamma_\rho & (\theta\le\psi\le 2\pi-\theta), \\ -r^{1/2}(\sin\theta/2 - i\cos\theta/2), \text{ if } \lambda = re^{i\theta} \in \Gamma_+ & (\rho\le r<\infty), \\ -r^{1/2}(\sin\theta/2 + i\cos\theta/2), \text{ if } \lambda = re^{-i\theta} \in \Gamma_- & (\rho\le r<\infty). \end{array} \right\} \quad (3.3)$$

Next, in order to avoid making the treatment too ponderous, we define certain integrals corresponding to the arcs Γ_ρ, Γ_+ and Γ_-:

$$I_\rho(y;\mu) := (2\pi i)^{-1} \int_{\Gamma_\rho} e^{y\sqrt{-\lambda}}\, R'(\lambda;A)\, (\sqrt{-\lambda})^{\mu}\lambda^{-1}d\lambda \quad (o<y), \qquad (3.4)$$

$$I_\pm(y;\mu) := (2\pi i)^{-1} \int_{\Gamma_\pm} e^{y\sqrt{-\lambda}}\, R'(\lambda;A)\, (\sqrt{-\lambda})^{\mu}\lambda^{-1}d\lambda \quad (o<y). \qquad (3.5)$$

Here μ denotes an arbitrary integer.

The following proposition precedes the main argument needed for the proof of Theorem (3.A).

PROPOSITION (3.B). Let the Sector condition (H1) be satisfied, and suppose that the Root definition (H2) holds. Then we have

(i) $I_\rho(\cdot;\mu)$, $I_\pm(\cdot;\mu) \in C(\mathbb{R}_+;B(E))$ for any integer μ,

(ii) $I_\rho(\cdot;\mu)$, $I_\pm(\cdot;\mu) \in L^p(\mathbb{R}_+;B(E)) \cap C_o(\mathbb{R}_+;B(E))$ for any $p \in [1,\infty)$ and any integer $\mu \le o$,

(iii) $I_\rho(\cdot;\mu)$, $I_\pm(\cdot;\mu) \in L^\infty(\mathbb{R}_+;B(E)) \cap C_o(\overline{\mathbb{R}}_+;B(E))$ for any integer $\mu \le -1$.

<u>Proof.</u> Let us first consider the integral $I_\rho(y;\mu)$. Using Corollary (2.N), we may find a constant $C>o$ not depending upon y and μ such that $\|I_\rho(y;\mu)\| \le C\rho^{\mu/2}\, e^{-y\sqrt{\rho}\,\sin\theta/2}$. From this inequality it follows that $I_\rho(y;\mu) \in B(E)$ $(y \ge o)$, and $\lim_{y\to+\infty} \|I_\rho(y;\mu)\| = o$ as well as

$$\int_{0}^{\infty} ||I_\rho(y;\mu)||^p \, dy < \infty \quad (1 \le p < \infty). \tag{3.6}$$

In order to prove continuity we let $y, y_o \in \overline{\mathbb{R}}_+$ be fixed and assume without any loss of generality that $y > y_o$. Then

$$||I_\rho(y;\mu) - I_\rho(y_o;\mu)|| \le C\rho^{\mu/2} e^{-y_o\sqrt{\rho} \sin\theta/2} \int_{\Gamma_\rho} |1-e^{(y-y_o)\sqrt{-\lambda}}| \, |d\lambda|.$$

Passing to the limit $y \to y_o$ we now obtain by a simple application of the Lebesgue Dominated-Convergence theorem the continuity at $y = y_o$. Since y_o was arbitrary, $I_\rho(y;\mu)$ is continuous everywhere. In particular, $I_\rho(\cdot;\mu)$ is strongly measurable by Pettis' theorem. This together with (3.6) imply that $I_\rho(\cdot;\mu) \in L^p(\mathbb{R}_+;B(E))$ $(1 \le p \le \infty)$ which concludes the proof for $I_\rho(y;\mu)$.

Next we consider the remaining integrals. Let C be chosen as before. The crucial point of the proof consists in establishing the estimates

$$||I_\pm(y;\mu)|| \le C \int_\rho^\infty e^{-y\sqrt{r} \sin\theta/2} \, r^{-1+\mu/2} \, dr$$

$$\le \begin{cases} C\mu^{-1}\rho^{\mu/2} e^{-y\sqrt{\rho} \sin\theta/2} & (\mu \le -1), \\ C(y\sqrt{\rho} \sin\theta/2)^{-1} e^{-y\sqrt{\rho} \sin\theta/2} & (\mu = o), \\ C(y \sin\theta/2)^{-\mu} \, 2\Gamma(\mu) & (\mu \ge 1). \end{cases}$$

Here $\Gamma(\mu)$ denotes the Eulerian Gamma function. Now we can repeat the arguments used in the preceding part of the proof. Difficulties may only arise in case of $\mu = o$ where we have to show the existence of the integral

$$J := \int_0^\infty \left(\int_{\sqrt{\rho}}^\infty e^{-yr \sin\theta/2}\, r^{-1}\, dr\right)^p dy \quad (1 \le p < \infty).$$

But an application of the Hölder inequality yields

$$J \le \int_0^\infty \int_{\sqrt{\rho}}^\infty e^{-pyr \sin\theta/2}\, r^{-1/2}\, dr\, dy \left(\int_{\sqrt{\rho}}^\infty r^{-(q+1)/2}\, dr\right)^{p/q}$$
$$(p^{-1} + q^{-1} = 1).$$

Thus, integrating first with respect to y and, applying Fubini's theorem, we will see that J exists. This concludes the proof of Proposition (3.B).

Proof of Theorem (3.A.i). Note that we have $S_o(y) = I_\rho(y;o) + I_+(y;o) + I_-(y;o)$. Thus $S_o(\cdot) \in L^p(\mathbb{R}_+;B(E)) \cap C_o(\mathbb{R}_+;B(E))$ is immediately obvious by statement (ii) of Proposition (3.B). In order to prove the continuity property $S_o(\cdot) \in C^\infty(\mathbb{R}_+;B(E))$ it suffices to show that $S_o(\cdot) \in C^\mu(\mathbb{R}_+;B(E))$ for any $\mu \in \mathbb{N}$. But using standard arguments for the justification of differentiation under the integral sign, we see that

$$\partial_y^\mu S_o(y) = I_\rho(y;\mu) + I_+(y;\mu) + I_-(y;\mu), \tag{3.7}$$

and hence the assertion follows from Proposition (3.B.i). This completes the proof.

Next we change to the proof of statement (ii) of Theorem (3.A). To this end we shall begin with a rather technical result.

LEMMA (3.C). Assume that the Root definition (H2) holds, and denote

$$J(y;\mu) := (2\pi i)^{-1} \int_L e^{y\sqrt{-\lambda}} (\sqrt{-\lambda})^{-\mu} \lambda^{-1} d\lambda.$$

Then it follows that $J(y;\mu) = o$ for all $y>o$ and any $\mu \in \mathbb{N} \cup \{o\}$. In case of $\mu>o$ this is even true for $y=o$.

<u>Proof.</u> We apply Cauchy's theorem in order to obtain

$$(2\pi i)^{-1} \int_{C_R} e^{y\sqrt{-\lambda}} (\sqrt{-\lambda})^{-\mu} \lambda^{-1} d\lambda = o$$

where $C_R \subset \mathbb{C}$ denotes the closed contour displayed in Fig. 2.

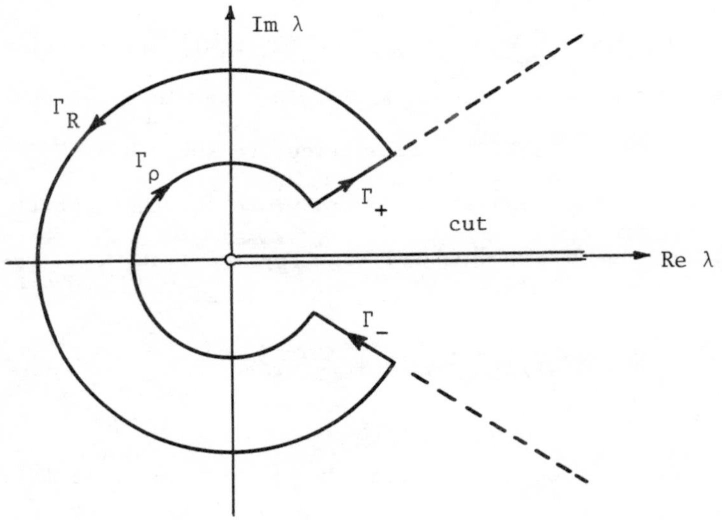

Fig.2

Thus

$$J(y;\mu) = -(2\pi i)^{-1} \lim_{R\to\infty} \int_{\Gamma_R} e^{y\sqrt{-\lambda}} (\sqrt{-\lambda})^{-\mu} \lambda^{-1} d\lambda$$

$$= -(2\pi)^{-1} \lim_{R\to\infty} R^{-\mu/2} \int_{\theta}^{2\pi-\theta} e^{-y\sqrt{R}(\sin\psi/2 - i\cos\psi/2)} (+i)^{-\mu} e^{-i\mu\psi/2} d\psi$$

$$= o,$$

as claimed.

Proof of Theorem (3.A.ii). Let $y>o$ be fixed. Using (3.7) and Proposition (2.B) we deduce that

$$A\,\partial_y^2\, S_o(y) = -(2\pi i)^{-1} \int_L e^{y\sqrt{-\lambda}}\, R'(\lambda;A)A d\lambda = -J(y;o) - S_o(y)$$

because the bounded operator A commutes with the integral sign. Now we refer to Lemma (3.C) to complete the proof.

In order to prove the remaining statement (iii) of Theorem (3.A) we need the following lemma which is also of interest for later use.

LEMMA (3.D). Let the Sector condition (H1) be satisfied, and suppose that the Root definition (H2) holds. Then we have for any $v \in E_o^1$

$$v = (2\pi i)^{-1} \int_L R'(\lambda;A)v\, \lambda^{-1}\, d\lambda = (2\pi i)^{-1} \int_L R'(\lambda;A)A^{-1}v\, \lambda^{-2}\, d\lambda$$

Proof. This lemma is a consequence of Cauchy's theorem of Residues. Indeed, let $v \in E_o^1$ be fixed. Since $A^{-1}: E_o^1 \to E$ is a bijection we may deduce from the identity (2.24) that

$$(2\pi i)^{-1} \int_L R'(\lambda;A)v\, \lambda^{-1} d\lambda = -A^{-1}v\, J(o;2) + (2\pi i)^{-1} \int_L R'(\lambda;A)A^{-1}v\, \lambda^{-2} d\lambda.$$

Note that $J(o;2) = o$ because of Lemma (3.C). Now let $C_R \subset \mathbb{C}$ denote the closed contour of Fig. 3

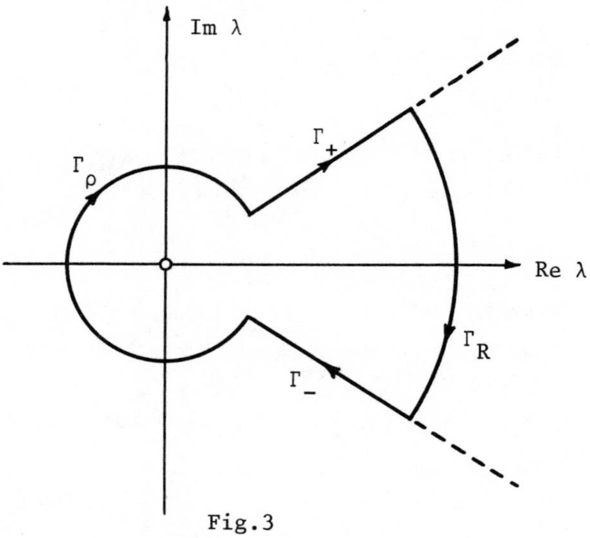

Fig.3

Integration along C_R yields

$$(2\pi i)^{-1} \int_{C_R} R'(\lambda;A)A^{-1}v\, \lambda^{-2}d\lambda = - \operatorname*{Res}_{\lambda=o} \lambda^{-2}R'(\lambda;A)A^{-1}v = v \tag{3.8}$$

for all $R>\rho>o$. Passing to the limit $R\to\infty$ it follows that

$$v = (2\pi i)^{-1} \int_L R'(\lambda;A)A^{-1}v\, \lambda^{-2}d\lambda$$
$$+ (2\pi i)^{-1} \lim_{R\to\infty} R^{-1} \int_\theta^{-\theta} R'(Re^{i\psi};A)A^{-1}ve^{-i\psi}d\psi,$$

and hence the lemma is proved because the last integral vanishes.

<u>Proof of Theorem</u> (3.A.iii). Let v be any element in E_o^1. Making use of

the identity (2.24) we may deduce from Lemma (3.C) that

$$S_o(y)v = (2\pi i)^{-1} \int_L e^{y\sqrt{-\lambda}} R'(\lambda;A)A^{-1}v\, \lambda^{-2} d\lambda$$

for all $y \geq o$. Combining this with the result of Lemma (3.D), we see that

$$\lim_{y\to o+} ||S_o(y)v-v|| \leq C\,||A^{-1}v||\, \lim_{y\to o+} \int_L |e^{y\sqrt{-\lambda}} -1|\; |\lambda^{-2}|\; |d\lambda|$$

with $C>o$ being a suitable constant. To conclude the proof we remark that the Dominated-Convergence theorem of Lebesgue applies to the last integral.

REMARK. Let $u_o \in E_o^1$ be arbitrary but fixed. In view of Theorem (3.A) it is clear that the function $u(y) = S_o(y)u_o$ is a classical solution of the two-point boundary value problem (3.2).

Recalling the results of Section 2.5, we are motivated to ask for an extension of Theorem (3.A.iii) to functions v belonging to E^1. Indeed, the assertions (3.A.iii) can be strengthened to the following

COROLLARY (3.E). Let the Sector condition (H1) be satisfied, and suppose that the Root definition (H2) holds. Then, if $E = L_+^p$ $(1 \leq p < \infty)$, it follows that $\lim_{y\to o+} ||S_o(y)v-v||_p = o$ for every $v \in E^1$. If in addition $k_o(t) = o$ $(t>o)$, and if $E \subseteq L_+^\infty$ then it follows that
$\lim_{y\to o+}\ \operatorname{ess.\,sup}_{t\geq t_o} |S_o(y)v(t) - v(t)| = o$ for every $v \in E^1$, where $t_o > o$ is arbitrary. If $E \nsubseteq L_+^\infty$ then the ess. sup may be simply replaced by sup.

<u>Proof.</u> We set $g(t) = e^{-kt}$ $(t \geq o)$ and $g(t) = o$ $(t<o)$. Let $v \in E^1$ be

arbitrary but fixed. Since $v(t) - v(o)g(t)$ belongs to E_o^1, it suffices to prove the corollary only for the function $v(t) = g(t)$. Thus consider the integral $(2\pi i)^{-1} \int_{C_R} R'(\lambda;A)g(t)\lambda^{-1}d\lambda$ where $C_R \subset \mathbb{C}$ is the closed contour of Fig. 3. Using Cauchy's theorem of residues combined with the results of Theorem (2.Q) and of Theorem (2.S), it is easy to see that $g(t) = (2\pi i)^{-1} \int_L R'(\lambda;A)g(t)\lambda^{-1}d\lambda$ $(t \geq o)$. Consequently, in case of $E = L_+^p$ $(1 \leq p < \infty)$ we infer that

$$\lim_{y\to o+} ||S_o(y)g-g||_p \leq C \lim_{y\to o+} \int_L |e^{y\sqrt{-\lambda}} -1| \; |\lambda|^{-(1+1/p)} \; |d\lambda|,$$

where C is a suitable constant determined by inequality (2.28). Equivalently, in case of $E \subseteq L_+^\infty$ we see that

$$\lim_{y\to o+} \operatorname{ess.\,sup}_{t\geq t_o} |S_o(y)g(t) - g(t)|$$

$$\leq \lim_{y\to o+} \int_L |e^{y\sqrt{-\lambda}} -1| \; |\lambda|^{-1} \; e^{-(k+|c||\lambda|(q-q_o))t_o} |d\lambda|,$$

where the results of Theorem (2.S) have been used. Apply now the Lebesgue Dominated-Convergence theorem in order to complete the proof.

We conclude this section with the observation that the relation $S_o(y)v = -A\,\partial_y^2\, S_o(y)v$ proved in Theorem (3.A) leads to

$$S_o(y)v \in E_o^1 \text{ for any } v \in E \text{ and each } y > o. \tag{3.9}$$

This implies that

$$\lim_{t\to o+} |S_o(y)v(t)| = o \quad (y > o) \tag{3.10}$$

because the functions of E_o^1 are absolutely continuous.

3.2. OPERATOR FAMILIES WHICH ARE RELATED TO $S_o(y)$

Notations of this section are the same as in the preceding section. Furthermore, δ_o will denote an arbitrary but fixed positive number.

In this section we change to the consideration of inhomogeneous two-point boundary value problems. In particular, we shall be concerned with the following two types

$$\left.\begin{aligned} &A\,\partial_y^2 u(y) + u(y) = Af(y) \quad (o < y < \infty), \\ &u(o) = u_o \quad \text{and} \quad \lim_{y\to+\infty} ||u(y)|| = o, \end{aligned}\right\} \tag{3.11}$$

$$\left.\begin{aligned} &A\,\partial_y^2 u(y) + u(y) = Af(y) \quad (o < y < \delta), \\ &u(o) = o = u(\delta), \end{aligned}\right\} \tag{3.12}$$

where $\delta \geq \delta_o$ is an arbitrary number. An existence theory for these problems will be presented in Sections 3.4 and 3.5. In the meantime we will collect some preliminary results concerning operator families $S_j(y)$ $(j = 1,2)$ which are closely related to $S_o(y)$. If L denotes the oriented path of Fig. 1, then we define $S_j(y)$ $(y \geq o)$ by the Dunford-Taylor integrals

$$S_1(y) := (2\pi i)^{-1} \int_L e^{y\sqrt{-\lambda}}\, R'(\lambda;A)(\lambda\sqrt{-\lambda})^{-1}\, d\lambda, \tag{3.13}$$

$$S_2(y) := (2\pi i)^{-1} \int_L e^{y\sqrt{-\lambda}}\, (e^{2\delta\sqrt{-\lambda}}-1)^{-1}\, R'(\lambda;A)(\lambda\sqrt{-\lambda})^{-1}\, d\lambda \tag{3.14}$$

Later, these families shall be used for the generation of solutions to the inhomogeneous equations (3.11 and 3.12).

Various properties of $S_1(y)$ and $S_2(y)$ corresponding to those of $S_0(y)$ are summarized in the following theorem.

THEOREM (3.F). Let the Sector condition (H1) be satisfied, and suppose that the Root definition (H2) holds. Then we have

(i) $S_j(\cdot) \in C^\infty(\mathbb{R}_+;B(E)) \cap L^p(\mathbb{R}_+;B(E)) \cap C_o(\overline{\mathbb{R}}_+;B(E))$ $(1 \leq p < \infty;$ $j = 1,2)$.

In addition, $||S_1(y)||_{B(E)}$ is bounded uniformly with respect to $y \geq o$, and $||S_2(y)||_{B(E)}$ also uniformly with respect to $\delta \geq \delta_o$.

(ii) $A\, \partial_y^2\, S_j(y) + S_j(y) = o$ for all $y > o$ $(j = 1,2)$,

(iii) $\partial_y\, S_1(y) = S_0(y)$ for $y > o$, and $S_2(2\delta+y) - S_2(y) = S_1(y)$ for $y \geq o$ and $\delta \geq \delta_o$.

<u>Proof.</u> We divide the proof into several steps.

(a) Recalling the definitions (3.4 and 3.5) of I_ρ respectively $I_\pm$, it is immediately obvious that $S_1(y) = I_\rho(y;-1) + I_+(y;-1) + I_-(y;-1)$. Therefore, we are in the situations of Proposition (3.B.ii and 3.B.iii). Hence, the statement (i) is true for $S_1(y)$. In order to prove this statement for $S_2(y)$ it suffices to show that $|e^{2\delta\sqrt{-\lambda}} -1|$ can be bounded from below by a positive constant not depending upon $\lambda \epsilon L$ and $\delta \geq \delta_o$. Indeed, let $\lambda = re^{i\psi}$ be an element of L such that $\rho \leq r < \infty$ and $\theta \leq \psi \leq 2\pi - \theta$. Then we have

$|e^{2\delta\sqrt{-\lambda}} - 1| \geq 1-e^{-2\delta\sqrt{r}\, \sin\psi/2} \geq 1-e^{-2\delta_o\sqrt{\rho}\, \sin\theta/2}$. This completes the proof of the assertion (i).

(b) Next we shall prove assertion (ii) for $S_1(y)$. Note that we have $J(y;1) = o$ by Lemma (3.C). Repeating arguments used in the proof of Theorem

(3.A.ii) we find that for $y>o$

$$A\,\partial_y^2\, S_1(y) = -(2\pi i)^{-1} \int_L e^{y\sqrt{-\lambda}}\, R'(\lambda;A)A(\sqrt{-\lambda})^{-1}d\lambda = -\,J(y;1) - S_1(y),$$

and hence the statement is proved for $S_1(y)$. Now, changing the object of Lemma (3.C) but not the proof we easily see that

$$\tilde{J}(y) := (2\pi i)^{-1} \int_L e^{y\sqrt{-\lambda}}\, (e^{2\delta\sqrt{-\lambda}} - 1)^{-1}\, (\lambda\sqrt{-\lambda})^{-1}\, d\lambda = o \qquad (y \geq o).$$

Consequently $A\,\partial_y^2\, S_2(y) = -\,\tilde{J}(y) - S_2(y) = -\,S_2(y)$ $(y>o)$, which concludes the proof of assertion (ii).

Finally, the assertion (iii) follows straight away from the definitions of $S_j(y)$, and thus we have proved the theorem.

Observe that by statement (i) of Theorem (3.F) the continuous linear operator $S_1(o)$ exists. Applying Proposition (2.B) to the definition (3.13), there results

$$S_1(o) = (2\pi i)^{-1} \int_L R'(\lambda;A)(\lambda\sqrt{-\lambda})^{-1}d\lambda = (2\pi i)^{-1}A^{-1} \int_L R(\lambda;A^{-1})(\lambda\sqrt{-\lambda})^{-1}\, d\lambda.$$

The last integral formally defines by operational calculus an analytic function of A^{-1}, namely

$$S_1(o) = (-A)^{1/2} \in B(E) \tag{3.15}$$

(see A.E. Taylor [44]). Indeed, the justification of this relation is given by the following lemma.

LEMMA (3.G). Let the Sector condition (H1) be satisfied, and suppose that the Root definition (H2) holds. Then we have $(S_1(o))^2 = -A \in B(E)$.

<u>Proof.</u> We first note that by substraction of the identities $\lambda R'(\lambda;A)R'(\lambda';A) + \lambda R'(\lambda;A) = \lambda\lambda' R'(\lambda;A)AR'(\lambda';A) = \lambda' R'(\lambda;A)R'(\lambda';A) + \lambda' R'(\lambda' A)$ the <u>resolvent equation</u> of the reduced resolvent is obtained

$$R'(\lambda;A)R'(\lambda';A) = \lambda'(\lambda-\lambda')^{-1}R'(\lambda';A) - \lambda(\lambda-\lambda')^{-1}R'(\lambda;A) \quad (\lambda\neq\lambda') \qquad (3.16)$$

Next we denote by $L' \subset \mathbb{C}$ another copy of the oriented path L, slightly shifted to the right (see Fig. 4).

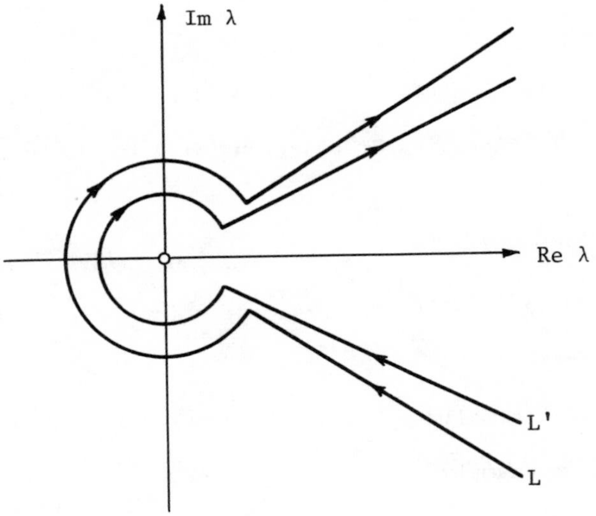

Fig.4

Since

$$(2\pi i)^{-1} \int_L R'(\lambda;A)(\lambda\sqrt{-\lambda})^{-1}d\lambda = S_1(o) = (2\pi i)^{-1} \int_{L'} R'(\lambda';A)(\lambda'\sqrt{-\lambda'})^{-1}d\lambda',$$

we may apply equation (3.16) and Fubini's theorem in order to see that

$$(S_1(o))^2 = (2\pi i)^{-2} \int_{L'} (\int_{L} [(\lambda-\lambda')\lambda\sqrt{-\lambda}]^{-1} d\lambda) R'(\lambda';A)(\sqrt{-\lambda'})^{-1} \, d\lambda'$$

$$- (2\pi i)^{-2} \int_{L} (\int_{L'} [(\lambda-\lambda')\lambda'\sqrt{-\lambda'}]^{-1} \, d\lambda') R'(\lambda;A)(\sqrt{-\lambda})^{-1} \, d\lambda.$$

Quite the same function theoretic arguments used in the proof of Lemma (3.C) imply that

$$(2\pi i)^{-1} \int_{L} [(\lambda-\lambda')\lambda\sqrt{-\lambda}]^{-1} d\lambda = o,$$

$$(2\pi i)^{-1} \int_{L'} [(\lambda-\lambda')\lambda'\sqrt{-\lambda'}]^{-1} d\lambda' = - (\lambda\sqrt{-\lambda})^{-1}.$$

From this and from (3.8) it follows that

$(S_1(o))^2 = - (2\pi i)^{-1} \int_{L} R'(\lambda;A) \, \lambda^{-2} d\lambda = -A$, and hence the proof is complete.

3.3. THE INFINITESIMAL GENERATOR OF THE OPERATOR FAMILY $S_o(y)$

The most famous semigroups of continuous linear operators are the <u>semigroups of class</u> (C_o) - thanks to E. Hille and K. Yosida who have stated the celebrated Hille-Yosida theorem. Such a semigroup is an one-parameter family of continuous linear operators, say $S(y) \in B(E)$ $(y \geq o)$, satisfying the conditions

$$S(y)S(y') = S(y+y') \quad \text{for} \quad y,y' \geq o, \tag{3.17}$$

$$S(o) = I \tag{3.18}$$

$$\lim_{y \to o+} ||S(y)v-v|| = o \quad \text{for each} \quad v \in E. \tag{3.19}$$

It is a well-known fact that each semigroup of class (C_o) is associated with a densely defined closed linear operator C which is defined by

$$D(C) := \{v \in E \mid \lim_{h \to o+} [S(h)v - v]/h \text{ exists in } E\},$$

$$Cv := \lim_{h \to o+} [S(h)v - v]/h \quad \text{for each } v \in D(C).$$

This operator is called the <u>infinitesimal</u> <u>generator</u> of $S(y)$. The importance of C with respect to the theory of differential equations in abstract spaces is manifested by the differential relation $\partial_y S(y)v = CS(y)v \quad (y \geq o)$ which holds for any $v \in D(C)$.

In this section we shall return to the operator family $S_o(y)$ of Section 3.1. We intend to show that there are quite a lot of familiarities between $S_o(y)$ and the semigroups of class (C_o), though $S_o(y)$ does <u>not</u> define such a semigroup. This is obvious, since by Theorem (3.A) the relation (3.19) only holds for each $v \in E_o^1$. However, we shall prove the semigroup property (3.17) - at least for $y,y' > o$ - and the existence of the infinitesimal generator.

THEOREM (3.H). Let the Sector condition (H1) be satisfied, and suppose that the Root definition (H2) holds. Then we have

(i) $S_o(y)S_o(y') = S_o(y+y')$ for $y,y' > o$.

(ii) $\lim_{h \to o+} [S_o(h)v - v]/h = -(2\pi i)^{-1} \int_L R'(\lambda;A)A^{-1}v(\lambda\sqrt{-\lambda})^{-1}d\lambda$ for each $v \in E_o^1$.

<u>Proof.</u> (i) Let $y,y' > o$ be arbitrary but fixed. Let L' be the oriented path of Fig. 4. We consider the operator product $S_o(y)S_o(y')$ where $S_o(y)$ and $S_o(y')$ may be defined respectively by the Dunford-Taylor integrals (3.1) taken on L and L'. We apply the resolvent equation (3.16) and obtain

$$S_o(y)S_o(y') = (2\pi i)^{-2} \int_{L'} (\int_L (\lambda-\lambda')^{-1}\lambda^{-1}e^{y\sqrt{-\lambda}}\, d\lambda)e^{y'\sqrt{-\lambda'}}R'(\lambda';A)d\lambda'$$

$$- (2\pi i)^{-2} \int_L (\int_{L'} [(\lambda-\lambda')\lambda']^{-1}\, e^{y'\sqrt{-\lambda'}}d\lambda')e^{y\sqrt{-\lambda}}R'(\lambda;A)d\lambda,$$

where we have made free use of Fubini's theorem. Following the proof of Lemma (3.G) we see that

$$(2\pi i)^{-1} \int_L (\lambda-\lambda')^{-1}\lambda^{-1}\, e^{y\sqrt{-\lambda}}d\lambda = 0,$$

$$(2\pi i)^{-1} \int_{L'} [(\lambda-\lambda')\lambda']^{-1}\, e^{y'\sqrt{-\lambda'}}d\lambda' = -\lambda^{-1}\, e^{y'\sqrt{-\lambda}},$$

and therefore we get

$S_o(y)S_o(y') = (2\pi i)^{-1} \int_L e^{y\sqrt{-\lambda}}\, e^{y'\sqrt{-\lambda}}\, R'(\lambda;A)\lambda^{-1}\, d\lambda = S_o(y+y')$ as claimed.

(ii) According to the identity (2.24) the function $S_o(y)v$ can be represented in the form $S_o(y)v = (2\pi i)^{-1} \int_L e^{y\sqrt{-\lambda}}\, R'(\lambda;A)A^{-1}v\, \lambda^{-2}d\lambda - A^{-1}v\, J(y;2)$, $(v \in E_o^1)$. But $J(y;2) = 0$ by Lemma (3.C). Thus in virtue of Lemma (3.D) we see that for $h>0$

$$\Delta_+S_o(h)v := [S_o(h)v-v]/h = (2\pi i)^{-1} \int_L h^{-1}(e^{h\sqrt{-\lambda}}-1)R'(\lambda;A)A^{-1}v\, \lambda^{-2}d\lambda.$$

Therefore

$$\Delta_+S_o(h)v + (2\pi i)^{-1} \int_L R'(\lambda;A)A^{-1}(\lambda\sqrt{-\lambda})^{-1}d\lambda$$

$$= (2\pi i)^{-1} \int_L [(e^{h\sqrt{-\lambda}} -1)h^{-1}- \sqrt{-\lambda}]\, R'(\lambda;A)A^{-1}v\, \lambda^{-2}d\lambda.$$

Passing to the limit $h\to 0+$ and applying the Dominated-Convergence theorem of Lebesgue we obtain the desired result from

$$\lim_{h\to o+} |(e^{h\sqrt{-\lambda}}-1)h^{-1} - \sqrt{-\lambda}| = o.$$

This completes the proof of Theorem (3.H).

REMARKS. 1°. Combining the above result on the infinitesimal generator with the result of Lemma (3.G) we infer that

$$\lim_{h\to o+} [S_o(h)v-v]/h = -(-A)^{1/2} A^{-1}v = (-A^{-1})^{1/2}v, \quad v \in E_o^1$$

where the last equality is valid at least in a formal sense. Thus the infinitesimal generator of the semigroup $S_o(y)$ $(y>o)$ is a densely defined closed operator which can be identified with the square root of the closed operator $-A^{-1}$.

2°. In some sense the statement (ii) of Theorem (3.H) can be extended to functions v belonging only to E^1. To be more precise, we have to strengthen the hypotheses of Theorem (3.H) to those of Corollary (3.E) and to take the limit $h\to o+$ in the sense of Corollary (3.E). We shall omit the details.

We close this section with the derivation of another representation formula for the semigroup $S_o(y)$. Since this formula consists of a single improper integral taken over the reals $\mathbb{R}_+$, it is more suitable for applications than the relatively complicated definition (3.1).

LEMMA (3.I). Let the Sector condition (H1) be satisfied, and suppose that the Root definition (H2) holds. Then $S_o(y)$ can be represented by

$$S_o(y) = I + \pi^{-1} \int_o^{\infty} \sin y\cdot\sqrt{r}\, R'(r;A)r^{-1}\, dr \quad (y > o). \tag{3.20}$$

Proof. Let $y>o$ be arbitrary but fixed. Let $\varepsilon>o$ and $R>o$ be any numbers satisfying $o < \varepsilon < \rho < R$. Here $\rho>o$ denotes the radius of the arc Γ_ρ. Let $C_R \subset \mathbb{C}$ be the closed contour of Fig. 5.

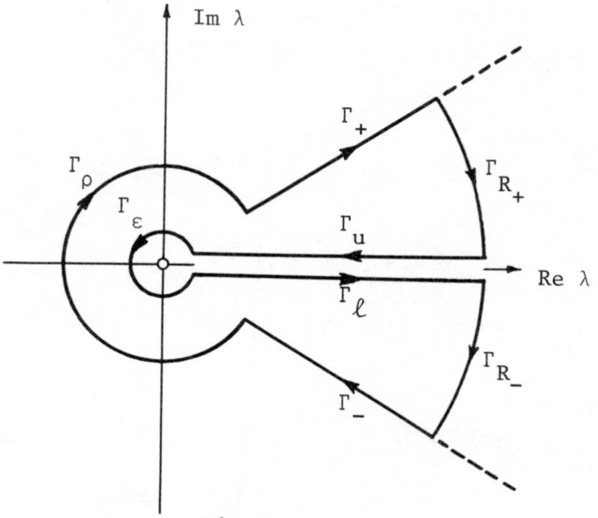

Fig.5

Then it follows from Cauchy's theorem that $(2\pi i)^{-1} \int\limits_{C_R} e^{y\sqrt{-\lambda}} R'(\lambda;A)\lambda^{-1} d\lambda = o$. Therefore,

$$(2\pi i)^{-1} \int\limits_{L} e^{y\sqrt{-\lambda}} R'(\lambda;A)\lambda^{-1} d\lambda = - \lim_{\substack{\varepsilon\to o+ \\ R\to+\infty}} \{ \int\limits_{\Gamma_{R_+}} + \int\limits_{\Gamma_{R_-}} + \int\limits_{\Gamma_u} + \int\limits_{\Gamma_\ell} + \int\limits_{\Gamma_\varepsilon} \}, \tag{3.21}$$

where in each integral appearing on the right-hand side the function $(2\pi i)^{-1} e^{y\sqrt{-\lambda}} R'(\lambda;A)\lambda^{-1}$ has to be inserted. We consider these integrals individually. To begin with, we substitute $\lambda = Re^{i\psi}$ in Γ_{R_+} and Γ_{R_-} such that

$$\| \int_{\Gamma_{R_+}} + \int_{\Gamma_{R_-}} \| \le C\left(\int_0^{\theta} e^{-y\sqrt{R}\,\sin\,\psi/2}\, d\psi + \int_{2\pi-\theta}^{2\pi} e^{-y\sqrt{R}\,\sin\,\psi/2}\, d\psi\right) \to o \quad (R\to\infty).$$

Next, in virtue of (2.23) a substitution $\lambda = \varepsilon e^{i\psi}$ in Γ_ε yields

$$\int_{\Gamma_\varepsilon} = (2\pi)^{-1} \int_0^{2\pi} e^{-y\sqrt{\varepsilon}\,(\sin\,\psi/2 \,-\, i\,\cos\,\psi/2)}\, R'(\varepsilon e^{i\psi};A)d\psi \to -I \quad (\varepsilon \to o+).$$

Finally, substituting $\lambda=r$ and $\lambda = re^{2\pi i}$ in Γ_u respectively in Γ_ℓ we obtain

$$\int_{\Gamma_u\cup\Gamma_\ell} = -\,\pi^{-1} \int_\varepsilon^R \frac{\sin\,y\sqrt{r}}{r}\, R'(r;A)dr.$$

In this last integral we can pass to the limit $\varepsilon \to o+$ without any difficulties, whereas the existence of the limit $\lim\limits_{R\to+\infty} \pi^{-1} \int_0^R \sin\,y\sqrt{r}\, R'(r;A)r^{-1}dr$ is a consequence of equation (3.21). This completes the proof.

REMARKS. 1°. It is important to remark that the above representation formula fails to be true for $y=o$ because in this case the proof of Lemma (3.I) is wrong.

2°. Representation formulae similar to (3.20) can also be derived for the operator families $S_j(y)$ $(j = 1,2)$. We do not need them, and therefore their derivation is left to the reader.

3.4. ABSTRACT TWO-POINT BOUNDARY VALUE PROBLEMS

Besides the notations of the previous section we will use the symbol $VB(\overline{\mathbb{R}_+};E)$ in order to denote the class of functions $f\colon \overline{\mathbb{R}_+} \to E$ which have <u>bounded variation</u> on every finite interval $[o,T]$, $o < T < \infty$.

It is well-known that any function f belonging to $VB(\overline{\mathbb{R}_+};E)$ admits function limits

$$f_+(y) := \lim_{h\to o+} f(y+h) \quad \text{and} \quad f_-(y) := \lim_{h\to o+} f(y-h) \quad (y \neq o) \tag{3.22}$$

in every point $y \in \overline{\mathbb{R}_+}$. Thus the only possible discontinuities of f are jumps of finite magnitude. Moreover, the set $\{y \in \mathbb{R}_+ \mid f_+(y) \neq f_-(y)\}$ is at most countable. Every point $y \in \overline{\mathbb{R}_+}$ is a (right or left) Lebesgue point of f which means that

$$f_+(y) = \lim_{h\to o+} h^{-1} \int_y^{y+h} f(z)dz \quad (y \in \overline{\mathbb{R}_+}),$$

$$f_-(y) = \lim_{h\to o-} h^{-1} \int_y^{y+h} f(z)dz \quad (y \in \mathbb{R}_+).$$

Furthermore, we shall denote by $B(X,E)$ the Banach algebra of all continuous linear operators mapping X into E where in general X is a normed complex space.

In this section we shall be concerned with the two-point boundary value problems (3.11 and 3.12) where the operator $A = GK_c$ is required to satisfy the Sector condition (H1). We shall begin with the treatment of problem (3.11). To this purpose we define a linear mapping R_1 acting on functions $f \in L^p(\mathbb{R}_+;E)$ $(1 \leq p \leq \infty)$ by means of the following Bochner integrals

$$R_1(y)f := \frac{1}{2}\left[\int_y^\infty S_1(z-y)f(z)dz + \int_o^y S_1(y-z)f(z)dz - \int_o^\infty S_1(y+z)f(z)dz\right]$$

$$(o \leq y < \infty). \tag{3.23}$$

Clearly, $S_1(y)$ denotes the operator family (3.13), and in view of Theorem

(3.F) the above definition makes sense.

The main result of this section is the following theorem which gives simultaneously a general existence result for the problem (3.11).

THEOREM (3.K). Let the Sector condition (H1) be satisfied, and suppose that the Root definition (H2) holds. Then we have

(i) $R_1(y) \in B(L^p(\mathbb{R}_+;E),E)$ for each $y \geq o$ and any fixed $1 \leq p \leq \infty$. $R_1(\cdot)f \in C(\overline{\mathbb{R}_+};E)$ and $\lim_{y\to o+} \|R_1(y)f\| = o$ for any $f \in L^p(\mathbb{R}_+;E)$ $(1 \leq p \leq \infty)$. If the limit $\lim_{y\to\infty} \|f(y)\| = o$ exists, then we also have $\lim_{y\to\infty} \|R_1(y)f\| = o$.

(ii) $\partial_y R_1(y) \in B(L^p(\mathbb{R}_+;E) \cap VB(\overline{\mathbb{R}_+};E),E)$ for each $y \geq o$ and any fixed $1 < p \leq \infty$, where $L^p(\mathbb{R}_+;E) \cap VB(\overline{\mathbb{R}_+};E)$ may be endowed with the induced norm topology of $L^p(\mathbb{R}_+;E)$. $\partial_y R_1(\cdot)f \in C(\overline{\mathbb{R}_+};E)$ for any $f \in L^p(\mathbb{R}_+;E) \cap VB(\overline{\mathbb{R}_+};E)$ $(1 < p \leq \infty)$.

(iii) For any $f \in VB(\overline{\mathbb{R}_+};E)$ satisfying $f(y) \in E_o^1$ $(y \geq o)$ and $G^{-1}f \in L^p(\mathbb{R}_+;E)$ for at least one $p \in [1,\infty]$, we have that $R_1(\cdot)f \in C^1(\overline{\mathbb{R}_+};E)$. Furthermore,

$$\left.\begin{aligned} A\partial_y^+ R_1'(y)f + R_1(y)f &= Af_+(y) \quad \text{for each } y \geq o \\ A\partial_y^- R_1'(y)f + R_1(y)f &= Af_-(y) \quad \text{for each } y > o \end{aligned}\right\} \quad (3.24)$$

where $\partial_y^\pm R_1'(y)f$ denotes respectively the right and left derivatives of $R_1'(y)f := \partial_y R_1(y)f$. The set $N := \{y>o \mid f_+(y) \neq f_-(y)\}$ is at most countable, and it follows that

$$A\partial_y^2 R_1(y)f + R_1(y)f = Af(y) \quad \text{for every } y \in \mathbb{R}_+\backslash N. \quad (3.25)$$

Finally, we have $R_1(y)f \in E_o^1$ for each $y \geq o$.

The proof of this theorem is more involved, and therefore we shall divide it into several steps.

(a) First we shall prove the assertion (i). Suppose that f belongs to $L^p(\mathbb{R}_+;E)$ for a fixed $1 < p < \infty$. Then it follows from the Hölder inequality and from Theorem (3.F) that

$$||R_1(y)f|| \le C(\int_o^\infty ||S_1(z)||^q dz)^{1/q} (\int_o^\infty ||f(z)||^p dz)^{1/p} < \infty \quad (p^{-1} + q^{-1} = 1)$$

for each $y \ge o$, where C is a suitable constant. Clearly similar estimates are valid in case of f belonging to $L^1(\mathbb{R}_+;E)$ or to $L^\infty(\mathbb{R}_+;E)$. This shows that $R_1(y) \in B(L^p(\mathbb{R}_+;E),E)$. Moreover, the continuity of $R_1(\cdot)f$: $\overline{\mathbb{R}_+} \to E$ can be easily deduced from the property $S_1(\cdot) \in C(\overline{\mathbb{R}_+};B(E))$ (see Theorem (3.F)) and from the remark that any Bochner integral defines an absolutely continuous function. Consequently, $\lim_{y\to o+} ||R_1(y)f - R_1(o)f|| = o$ and, hence $\lim_{y\to o+} ||R_1(y)f|| = o$ since we have $R_1(o)f = o$.

Now we shall prove that $\lim_{y\to\infty} ||R_1(y)f|| = o$. Note that by hypothesis we have $\lim_{y\to\infty} ||f(y)|| = o$, and from Theorem (3.F) we infer that $\lim_{y\to\infty} ||S_1(y)|| = o$. The first relation leads to

$$\lim_{y\to\infty} ||\int_y^\infty S_1(z-y)f(z)dz|| \le \int_o^\infty ||S_1(z)|| \lim_{y\to\infty} ||f(y+z)||dz = o,$$

and similarly the second relation yields

$$\lim_{y\to\infty} ||\int_o^\infty S_1(y+z)f(z)dz|| \le \int_o^\infty \lim_{y\to\infty} ||S_1(y+z)|| \, ||f(z)||dz = o.$$

In both cases we may justify the interchange of integration and passing to limits by the Dominated-Convergence theorem of Lebesgue. Now it remains to prove that

$$\lim_{y\to\infty} \|\int_0^y S_1(y-z)f(z)dz\| = o. \tag{3.26}$$

Indeed by assumption we may determine for any $\varepsilon>o$ a number $N(\varepsilon) > o$ such that $\|f(y)\| \le \varepsilon$ holds for all $y > N(\varepsilon)$. Assuming that $y > N(\varepsilon)$ we obtain

$$\|\int_0^y S_1(y-z)f(z)dz\| \le \int_0^{N(\varepsilon)} \|S_1(y-z)\| \, \|f(z)\|dz + \varepsilon \int_{N(\varepsilon)}^y \|S_1(y-z)\|dz.$$

Passing to the limit $y\to\infty$, clearly the integral in the middle vanishes, whereas the last integral is bounded by $\varepsilon \int_0^\infty \|S_1(z)\|dz$. Since $\varepsilon>o$ was arbitrarily chosen we obtain $\overline{\lim}_{y\to\infty} \|\int_0^y S_1(y-z)f(z)dz\| = o$, and this implies the assertion (3.26).

(b) Next we shall prove the relation

$$\partial_y R_1(y)f = \frac{1}{2}\,[-\int_y^\infty S_o(z-y)f(z)dz + \int_0^y S_o(y-z)f(z)dz - \int_0^\infty S_o(y+z)f(z)dz]$$

$$(y \ge o), \tag{3.27}$$

where f is assumed to belong to $L^p(\mathbb{R}_+;E) \cap VB(\overline{\mathbb{R}_+};E)$ for a fixed $1 < p \le \infty$. We begin the proof with the computation of the right derivative $\partial_y^+ R_1(y)f$. To this purpose let $y \ge o$ be fixed. Consider for the moment only the integral $\tilde{R}_1(y)f := \int_y^\infty S_1(z-y)f(z)dz$. Then

$$\partial_y^+ \tilde{R}_1(y)f = \lim_{h\to o+} h^{-1}(\tilde{R}_1(y+h)f - \tilde{R}_1(y)f)$$

$$= \lim_{h\to o+} \{\int_{y+h}^\infty [S_1(z-y-h) - S_1(z-y)]h^{-1}f(z)dz - h^{-1}\int_y^{y+h} S_1(z-y)f(z)dz\}.$$

Following a standard device we see that the first integral converges to

$$- \int_y^\infty \partial_z S_1(z-y)f(z)dz = - \int_y^\infty S_0(z-y)f(z)dz,$$

where the equality is a consequence of Theorem (3.F.iii). As for the second integral we define $F(\xi) := \int_y^{y+\xi} S_1(z-y)f(z)dz \quad (\xi \geq 0)$. Clearly, F is strongly differentiable. Therefore we have

$$\partial_\xi^+ F(0) = S_1(0)f_+(y) = \lim_{h \to 0+} [F(h)-F(0)]/h = \lim_{h \to 0+} h^{-1} \int_y^{y+h} S_1(z-y)f(z)dz.$$

Summarizing to this point we just have proved that

$$\partial_y^+ \tilde{R}_1(y)f = - S_1(0)f_+(y) - \int_y^\infty S_0(z-y)f(z)dz \qquad (y \geq 0).$$

In quite the same way we also may prove that

$$\partial_y^+ \int_0^y S_1(y-z)f(z)dz = S_1(0)f_+(y) + \int_0^y S_0(y-z)f(z)dz \qquad (y \geq 0),$$

$$\partial_y^+ \int_0^\infty S_1(y+z)f(z)dz = \int_0^\infty S_0(y+z)f(z)dz \qquad (y \geq 0).$$

Corresponding results, with $f_+(y)$ replaced by $f_-(y)$, can be derived for the left derivative $\partial_y^- R_1(y)f$. Collecting these results we thus have proved the relation (3.27).

We are now in the position to complete the proof of the assertion (ii). Recall that $S_0(\cdot)$ belongs to $L^q(\mathbb{R}_+;B(E))$ for all $1 \leq q < \infty$ (see Theorem (3.A)). Thus we may simply follow the argumentation used in the preceding point (a) in order to deduce from (3.27) the desired properties of $\partial_y R_1(y)f$.

(c) We conclude the proof of Theorem (3.K) by proving statement (iii). By assumption, we have $G^{-1}f \in L^p(\mathbb{R}_+;E)$ which implies that $A^{-1}f \in L^p(\mathbb{R}_+;E)$. Recalling the identity (2.24) we may establish the relation (3.27) under the modified hypotheses of statement (iii). Then we may differentiate the relation (3.27) by repeating essentially the calculations of the preceding point (b). This leads to

$$\left.\begin{aligned} \partial_y^{(\pm)} R_1'(y)f = f_{(\pm)}(y) + \frac{1}{2}\{\int_y^\infty S_o'(z-y)f(z)dz + \int_o^y S_o'(y-z)f(z)dz\} \\ - \frac{1}{2}\int_o^\infty S_o'(y+z)f(z)dz, \quad y \geq o \quad (y > o), \end{aligned}\right\} \tag{3.28}$$

where we have used the notation $S_o'(y)$ for $\partial_y S_o(y) = \partial_y^2 S_1(y)$. Observing that $A\,\partial_y^2 S_1(y) = -S_1(y)$ (see Theorem (3.F)) we immediately obtain the relations (3.24). Since the remaining assertions are easy consequences of (3.24), the proof is complete.

REMARKS. 1°. Let f satisfy the assumptions of Theorem (3.K.iii). Then a combination of Theorem (3.A) together with Theorem (3.K) yields that the function $u(y) = S_o(y)u_o + R_1(y)f$ is a solution of the two-point boundary value problem (3.11), provided that $u_o \in E_o^1$. The question of uniqueness will be treated in chapter 5.

2°. The results of Theorem (3.K.iii) can be extended to functions $f \in VB(\overline{\mathbb{R}_+};E)$ satisfying the conditions $f(y) \in E^1$ $(y \geq o)$ and $(k+\partial_t)f \in L^p(\mathbb{R}_+;E)$. However in this case we have to strengthen the hypotheses of the theorem to those of Corollary (3.E). Another result in this direction will be derived in the next section where we consider Hölder continuous functions f.

We shall continue our considerations on two-point boundary value problems by changing over to problem (3.12). This problem looks very much like the problem (3.11). Indeed, its analysis does not differ too much from the analysis developed in the preceding part of this section.

For the rest of this section let us assume that $\delta_o > o$ is a fixed number, and that $\delta \geq \delta_o$ can be arbitrarily chosen. Furthermore, replacing $\mathbb{R}_+$ by (o,δ) and $\overline{\mathbb{R}_+}$ by $[o,\delta]$, we change the notations of function spaces based on $\mathbb{R}_+$ in an obvious manner.

Corresponding to the approach accomplished for the problem (3.11), we will define a linear mapping R_2 acting on functions $f \in L^p(o,\delta;E)$ $(1 \leq p \leq \infty)$ in the following way

$$R_2(y)f := \frac{1}{2}\int_o^\delta [S_2(y+z) - S_2(2\delta+y-z) + S_2(2\delta-y-z) - S_2(2\delta-y+z)]f(z)dz$$
$$+ \frac{1}{2}\{\int_o^y S_1(y-z)f(z)dz + \int_y^\delta S_1(z-y)f(z)dz\} \quad (o \leq y \leq \delta) \tag{3.29}$$

Here $S_1(y)$ and $S_2(y)$ are the operator families introduced by (3.13 and 3.14) respectively. In view of Theorem (3.F), each of the above Bochner integrals is well defined. In particular, property $S_2(2\delta+y) - S_2(y) = S_1(y)$ implies that

$$R_2(o)f = o = R_2(\delta)f. \tag{3.30}$$

Other properties of $R_2(y)$, including a general existence result for the problem (3.12), are obtained in much the same way as in the preceding case. We conclude this section by the following theorem which is a synopsis of these results.

THEOREM (3.L). Let the Sector condition (H1) be satisfied, and suppose that the Root definition (H2) holds. Then we have

(i) $R_2(y) \in B(L^p(o,\delta;E),E)$ for each $y \in [o,\delta]$ and any fixed $1 \le p \le \infty$, having a bound independent of $\delta \ge \delta_o$. $R_2(\cdot)f \in C([o,\delta];E)$ and $\lim_{y\to o+} ||R_2(y)f|| = o = \lim_{y\to\delta-} ||R_2(y)f||$ for any f belonging to $L^p(o,\delta;E)$ $(1 \le p \le \infty)$.

(ii) $\partial_y R_2(y) \in B(L^p(o,\delta;E) \cap VB([o,\delta];E),E)$ for each $y \in [o,\delta]$ and any fixed $1 < p \le \infty$. $\partial_y R_2(\cdot)f \in C([o,\delta];E)$ for any $f \in L^p(o,\delta;E) \cap VB([o,\delta];E)$ $(1 < p \le \infty)$.

(iii) For any $f \in VB([o,\delta];E)$ satisfying $f(y) \in E_o^1$ $(o \le y \le \delta)$ and $G^{-1}f \in L^p(o,\delta;E)$ for at least one $p \in [1,\infty]$, we have that $R_2(\cdot)f \in C^1([o,\delta];E)$. Furthermore,

$$\left.\begin{aligned} &A\, \partial_y^+ R_2'(y)f + R_2(y)f = Af_+(y) \quad \text{for each} \quad o \le y < \delta, \\ &A\, \partial_y^- R_2'(y)f + R_2(y)f = Af_-(y) \quad \text{for each} \quad o < y \le \delta, \end{aligned}\right\} \tag{3.31}$$

where $\partial_y^{\pm} R_2'(y)f$ denotes respectively the right and left derivatives of $R_2'(y)f := \partial_y R_2(y)f$. The set $N := \{y \in (o,\delta) \mid f_+(y) \neq f_-(y)\}$ is at most countable, and it follows that

$$A\, \partial_y^2 R_2(y)f + R_2(y)f = Af(y) \quad \text{for every} \quad y \in (o,\delta)\backslash N. \tag{3.32}$$

Finally, we have $R_2(y)f \in E_o^1$ for each $y \in [o,\delta]$.

3.5. HÖLDER CONDITIONS ON f(y)

In this section, we shall present another approach to the solution of the two-point boundary value problems considered in the foregoing part of this chapter. Notations will be the same as in Section 3.4.

We are motivated to give this modified approach by the following observations. In view of the circumstances associated with the singular perturbation problem (P_ε), Theorem (3.K) (and equivalently Theorem (3.L)) may not be applicable to it because of its incompatible hypotheses. In particular, great difficulties will arise in realizing the assumptions on f, namely the requirement of $f(y) \in E_o^1$ $(y \geq o)$, and of $G^{-1}f \in L^p(\mathbb{R}_+;E)$. As we have already remarked in Section 3.4, the first assumption on f may be weakened to that of $f(y) \in E^1$ $(y \geq o)$. Nevertheless, in case of $E \subseteq L_+^\infty$ we have to impose an additional condition on the kernel function $k_o(t)$. However, the second assumption remains still to be crucial.

From this we are forced to look for other suitable conditions on f. In this situation, it will turn out that the most appropriate class of functions is the class of <u>locally uniformly Hölder continuous functions</u> which are defined on $\mathbb{R}_+$ and which assume values in the Banach space E. To be more precise, these functions are required to satisfy the following condition.

(F) There exists a number $\alpha \in (o,1)$ with the property: For each fixed $\eta > o$ there is a constant $H = H(\eta)$ such that

$$||f(y) - f(z)|| \leq H(\eta)|y-z|^\alpha \quad \text{for every} \quad y,z \geq \eta.$$

Our principal result which in some sense is a generalization of Theorem (3.K.iii), will be stated as follows.

THEOREM (3.M). Let the Sector condition (H1) be satisfied, and assume that the Root definition (H2) holds. Let $R_1(y)$ $(y \geq o)$ denote the operator family (3.23). Then, for any locally uniformly Hölder continuous function $f: \mathbb{R}_+ \to E$ satisfying the condition (F), and belonging to $L^p(\mathbb{R}_+;E)$ for at least one $p \in [1,\infty]$, we have

(i) $R_1(\cdot)f \in C(\overline{\mathbb{R}_+};E) \cap C^2(\mathbb{R}_+;E)$.

(ii) $A\,\partial_y^2 R_1(y)f + R_1(y)f = Af(y)$ for each $y > o$, where from this relation it follows that $R_1(y)f \in E_o^1$ for every fixed $y > o$.

Proof. In order to prove the theorem, we proceed in several steps. In the first step we again prove the formula (3.27) to be valid, even for functions f belonging to $L^1(\mathbb{R}_+;E)$, but not at $y = o$.

(a) Justification of formula (3.27). As we have already mentioned, every function f satisfying the Hölder condition (F) is locally absolutely continuous, and a fortiori, for each fixed $\eta > o$, an element of $VB([\eta,\infty);E)$. Using Theorem (3.K.ii), we see that $\partial_y R_1(\cdot)f \in C([\eta,\infty);E)$, which is valid for any $\eta > o$ and any f belonging to $L^p(\mathbb{R}_+;E)$ $(1 < p \leq \infty)$. This implies (3.27). Now we want to show that formula (3.27) can be extended to functions f belonging to $L^1(\mathbb{R}_+;E)$. To prove this, it suffices to show that each integral on the right-hand side of equality (3.27) exists. To begin with, we consider the last integral $\int_o^\infty S_o(y+z)f(z)dz$. From the proof of Proposition (3.B) we infer that

$$||S_o(y+z)|| \leq C(1+y^{-1})e^{-y\sqrt{\rho}\,\sin\theta/2} \quad \text{uniformly for } z \geq o \tag{3.33}$$

where C is a suitable constant independent of y. Thanks to $f \in L^1(\mathbb{R}_+;E)$, this implies the existence of the above integral for each $y > o$. Next, we consider the integral $J := \int_y^\infty S_o(z-y)f(z)dz$. Letting $y > o$ be fixed, we choose the number η such that $o < \eta < y$. Observing that $J = \int_y^\infty S_o(z-y)[f(z) - f(y)]dz + \int_o^\infty S_o(z)f(y)dz$, we may apply the condition (F) in order to obtain

$$||J|| \leq H(\eta) \int_o^\infty z^\alpha ||S_o(z)||dz + ||f(y)|| \int_o^\infty ||S_o(z)||dz.$$

In view of the estimate (3.33) and of the property $S_o \in L^1(\mathbb{R}_+;B(E))$ it is easily checked that the above integrals exist. Finally, we consider the integral $\tilde{J} := \int_o^y S_o(y-z)f(z)dz$. Again we let $o < \eta < y$. Then, by inequality (3.33) and by condition (F), we see that

$$||\tilde{J}|| \leq C[1+(y-\eta)^{-1}]e^{-(y-\eta)\sqrt{\rho}\,\sin\theta/2} \int_o^\eta ||f(z)||dz$$
$$+ ||f(y)|| \int_o^{y-\eta} ||S_o(z)||dz + H(\eta) \int_o^{y-\eta} z^\alpha ||S_o(z)||dz,$$

each integral being finite. This proves the formula (3.27).

(b) Second derivatives. Here we want to show the following relation to be valid for $y > o$:

$$\left.\begin{aligned} \partial_y^2 R_1(y)f &= \tfrac{1}{2} S_o(y)f(y) + \tfrac{1}{2} \int_y^\infty S_o'(z-y)[f(z) - f(y)]dz \\ &+ \tfrac{1}{2} \{\int_o^y S_o'(y-z)[f(z) - f(y)]dz - \int_o^\infty S_o'(y+z)f(z)dz\}, \end{aligned}\right\} \quad (3.34)$$

where $S_o'(y)$ denotes the strong derivative of $S_o(y)$, $S_o'(y) = \partial_y S_o(y) = \partial_y^2 S_1(y)$. To this end we let $y > o$ being arbitrary but fixed. Then we choose the number η such that $o < \eta < y$. For the moment we cosider only the integral $\tilde{R}_1'(y)f := \int_o^y S_o(y-z)f(z)dz$. If $h \neq o$ is any given number satisfying $\eta < y + h$, then we may introduce the difference quotient of $\tilde{R}_1'$

at $y > o$ with respect to f, namely

$$\Delta(y,f;h) := h^{-1}[\tilde{R}_1'(y+h)f - \tilde{R}_1'(y)f]$$

$$= h^{-1} \{ \int_0^{y+h} S_0(y+h-z)f(z)dz - \int_0^{y} S_0(y-z)f(z)dz \}.$$

The following identity can easily be checked.

$$\Delta(y,f;h) = h^{-1} \{ \int_0^{y+h} S_0(y+h-z)[f(z)-f(y)]dz - \int_0^{y} S_0(y-z)[f(z)-f(y)]dz \}$$

$$+ h^{-1} \{ \int_0^{y+h} S_0(y+h-z)f(y)dz - \int_0^{y} S_0(y-z)f(y)dz \}$$

$$=: J_1(h) + J_2(h).$$

Here we consider first the integral $J_2(h)$. After some rearrangements we obtain

$$J_2(h) = h^{-1} \int_y^{y+h} S_0(z)f(y)dz.$$

Since the mapping $z \mapsto S_0(z)f(y)$ is strongly continuous, every point $z>o$ is a Lebesgue point. Hence, we conclude that

$$\lim_{h\to o} J_2(h) = S_0(y)f(y) \quad \text{for each } y > o.$$

Next, we change over to the consideration of $J_1(h)$. Without any loss of generality we may assume that $h > o$. For $h < o$, the following calculations can also be carried out with minor changes. It is immediately obvious that

$$J_1(h) = \int_0^y h^{-1} [S_0(y+h-z) - S_0(y-z)][f(z)-f(y)]dz$$

$$+ h^{-1} \int_y^{y+h} S_0(y+h-z) [f(z)-f(y)]dz.$$

As for the last integral we may use the definitions (3.4 and 3.5), and the condition (F) in order to see that

$$||S_0(y+h-z)[f(z)-f(y)]||$$

$$\leq H(\eta)h^{\alpha} \{||I_{\rho}(y+h-z;o)|| + ||I_+(y+h-z;o)|| + ||I_-(y+h-z;o)||\}.$$

Integrating with respect to $z \in (y,y+h)$, and using the estimates for I_ρ and $I_\pm$ established in the proof of Proposition (3.B), this yields together with Fubini's theorem

$$||h^{-1} \int_y^{y+h} S_0(y+h-z)[f(z)-f(y)]dz|| \leq Ch^{\alpha-1}(1-e^{-h\sqrt{\rho}\sin\theta/2}).$$

Here C is a suitable constant not depending upon h. Passing now to the limit $h \to o+$ we see that the integral vanishes. Using quite the same arguments, one can also prove that

$$\lim_{h\to o+} \int_o^y h^{-1} [S_0(y+h-z) - S_0(y-z)][f(z)-f(y)]dz = \int_o^y S_0'(y-z)[f(z)-f(y)]dz.$$

Summarizing to this point, we just have proved that for $y > o$

$$\partial_y \tilde{R}_1'(y)f = \lim_{h\to o} \Delta(y,f;h) = S_0(y)f(y) + \int_o^y S_0'(y-z)[f(z)-f(y)]dz.$$

Now, by following the outlines of the above proof for the remaining integrals of formula (3.27), it is easily checked that

$$\partial_y \int_y^\infty S_o(z-y)f(z)dz = -\int_y^\infty S_o'(z-y)(f(z)-f(y))dz \qquad (y > o),$$

$$\partial_y \int_o^\infty S_o(y+z)f(z)dz = \int_o^\infty S_o'(y+z)f(z)dz \qquad (y > o).$$

Putting all these results into formula (3.27), we then obtain the relation (3.34).

(c) Conclusions. By hypothesis it follows that f belongs to $C(\mathbb{R}_+;E) \cap L^p(\mathbb{R}_+;E)$, and therefore the assertion (i) of Theorem (3.M) is an immediate consequence of Theorem (3.K.i) and of formulae (3.27 and 3.34). In order to prove the assertion (ii), we apply the operator A on both sides of (3.34). Observing that A commutes with the integral signs, and that $-AS_o'(y) = S_1(y)$ $(y > o)$, we infer that

$$A\partial_y^2 R_1(y)f = -R_1(y)f + \frac{1}{2}[AS_o(y)f(y) + \int_y^\infty S_1(z-y)f(y)dz - \int_o^y S_1(y-z)f(y)dz].$$

Thus, it remains to simplify the term within the brackets. But this can be rewritten as

$$2(2\pi i)^{-1} \int_L R'(\lambda;A)f(y)\lambda^{-2}d\lambda = 2Af(y) \qquad (y > o),$$

where the last equality is obtained from the relation (3.8), and where we have used the definitions of S_o and S_1 in connection with Fubini's theorem. This completes the proof of Theorem (3.M).

The corresponding result for the operator family R_2 can be obtained by using essentially the same proof technique. We conclude this section with the following theorem.

THEOREM (3.N). Let the Sector condition (H1) be satisfied, and suppose that the Root definition (H2) holds. Let $R_2(y)$ $(y \geq o)$ denote the operator family (3.29). Then for any locally uniformly Hölder continuous function $f: (o,\delta) \to E$ satisfying the condition (F), and belonging to $L^p(o,\delta;E)$ for at least one $p \in [1,\infty]$, we have

(i) $R_2(\cdot)f \in C([o,\delta];E) \cap C^2((o,\delta);E)$,

(ii) $A\partial_y^2 R_2(y)f + R_2(y)f = Af(y)$ for each $y \in (o,\delta)$, where from this relation it follows that $R_2(y)f \in E_o^1$ for every fixed $y \in (o,\delta)$.

BIBLIOGRAPHICAL NOTES

Sec. 3.1 and 3.2. For a general discussion of the theory of Dunford-Taylor integrals, the reader is referred to the treatises of A.E. Taylor [44], and of N. Dunford and J.T. Schwartz [13]. The representation of fractional powers of an operator by means of a Dunford-Taylor integral is essentially due to V. Balakrishnan [2]. The formula (3.15) occurs as a special case of a general representation formula obtained by Balakrishnan.

Sec. 3.3. The competent reference for all questions concerning semi-groups of continuous linear operators is the monograph of E. Hille and R.S. Phillips [26].

Sec. 3.4. The operator families $S_\mu(y)$ $(\mu = o,1,2)$ and $R_\mu(y)$ $(\mu = 1,2)$ have been introduced in the author's paper [23], where a simplified version of the Theorems (3.A) and (3.F) was already given. The idea of using functions of bounded variation for the right-hand sides of equations

(3.11 and 3.12) is adapted from the theory of nonlinear evolution equations, see T. Kato [33] and H. Brezis [4].

Sec. 3.5. The introduction of Hölder continuous functions was inspired by results which were obtained for inhomogeneous linear evolution equations involving holomorphic semigroups. We mention here the investigations of T. Kato [31, 32]. Other references will be found in R.W. Carroll [8].

4 The formal derivation of the asymptotic expansion

With the help of the preliminary analysis in chapters 2 and 3, we now propose a formal procedure for obtaining what are usually called the outer and inner expansions of the solution to the problem (P_ε). This procedure is based on the so called matching principle. A fundamental introduction into this principle and also its rigorous application to numerous problems of mathematical physics can be found in the book of W. Eckhaus [15].

To illustrate the idea, we shall describe in Section 4.1 the procedure by computing the first few terms of the formal outer expansion. Then we show that this procedure can be continued - by an induction argument - to be used for obtaining higher order terms.

In Section 4.2 we shall continue the description of the above procedure by establishing initial boundary value problems for the determination of the formal inner expansion. The main idea consists in introducing so called stretched variables. These variables magnify a suitable neighbourhood of the boundary unproportionally to the rest of the domain under consideration. In such a neighbourhood, to which it will be usually referred as the boundary layer region, the procedure of matching the outer and the inner expansions is accomplished.

Within the boundary layer region, the inner expansion dominates, and therefore it is also called the boundary layer or the boundary correction term. In the particular problem under consideration, we have something what in literature is called a parabolic boundary layer.

4.1. THE OUTER EXPANSION

In this section we shall begin with the derivation of the <u>outer expansion</u>. We shall use the notations of Section 2.3 for the continuous linear operators G, $K_c = cI + K_o$, $A = GK_c$, and by K_1 we denote the convolution operator generated by the kernel function $k_1 \in L^1(\mathbb{R})$,

$$K_1 v(t) := \int_0^\infty k_1(t-s) v(s) ds \quad (t \geq o).$$

For the present let Δ denote the Laplacian in n-dimensional Euclidean space $\mathbb{R}^n$, and let $\Omega \subset \mathbb{R}^n$ be any bounded open domain having a boundary $\partial\Omega$. Note that in the next section, we shall confine ourselves to the one-dimensional case.

Assume for a moment that the problem (P_ε) has a unique (classical) solution, $u = u(x,t;\varepsilon)$ $(x \in \overline{\Omega},\ t \geq o)$. An asymptotic expression for u is obtained formally by using the Taylor expansion about $\varepsilon = o$. To this purpose, let $m \geq o$ be any fixed integer. Let $U(x,t;\varepsilon)$ denote that part of the Taylor expansion which includes all terms up to the m-th order; then we may write U in the following form:

$$U(x,t;\varepsilon) = \sum_{\mu=o}^{m} \varepsilon^\mu U_\mu(x,t) \quad (x \in \overline{\Omega},\ t \geq o). \tag{4.1}$$

Clearly, U represents the solution u of (P_ε) asymptotically as $\varepsilon \to o+$. This is, by a substitution of (4.1) in (P_ε), asymptotically equivalent with the differential equation

$$\left.\begin{aligned} \sum_{\mu=o}^{m} \{\varepsilon^\mu (k+\partial_t) U_\mu(x,t) + \varepsilon^{\mu+1}[K_c \Delta U_\mu + K_1(k+\partial_t)U_\mu](x,t) &= \varepsilon r(x,t) \\ (x \in \Omega,\ t > o)& \end{aligned}\right\} \tag{4.2}$$

being subject to the initial condition

$$\sum_{\mu=o}^{m} \varepsilon^{\mu} U_{\mu}(x,o) = h(x) \quad (x\in\Omega), \tag{4.3}$$

and being subject to the boundary conditions which we do not consider here. We will come back to this point in the next section.

Comparing now same powers of ε in (4.2) and (4.3), we are led to the following initial value problems for the determination of U_{μ} $(\mu = o,1,\dots,m)$:

$$\begin{cases} (k+\partial_t)U_o(x,t) = o & (x\in\Omega,\ t > o), \\ U_o(x,o) = h(x) & (x\in\Omega), \end{cases} \tag{P_o}$$

$$\begin{cases} (k+\partial_t)U_{\mu}(x,t) = \delta_{1\mu} r(x,t) - [K_c \Delta U_{\mu-1} + K_1(k+\partial_t)U_{\mu-1}](x,t) & (x\in\Omega,\ t>o), \\ U_{\mu}(x,o) = o & (x\in\Omega). \end{cases} \tag{P_{μ}}$$

Here $\delta_{1\mu}$ stands for the usual Kronecker symbol.

On the other hand, let U_{μ} denote the solutions of the initial value problems (P_{μ}) $(\mu = o,1,\dots,m)$. By virtue of (4.1), we then may compose these solutions to the function $U(x,t;\varepsilon)$. Again, U represents clearly the solution u of (P_{ε}), asymptotically as $\varepsilon \to o+$, but at this time only for values x away from the boundary $\partial\Omega$, because the problems (P_{μ}) admit no further conditions to be imposed on U_{μ} at $\partial\Omega$. In other words, U represents asymptotically the solution u <u>outside</u> some boundary layer region. Therefore, to U it will be usually referred as the <u>outer expansion</u>.

At this stage of approximation, it would be important to introduce boundary correction terms, which should take care for the given boundary data.

However, this process is important enough to be treated separately in the next section. Thus, we shall continue here to consider the problems (P_μ). We are mainly interested in explicit representation formulae for the solutions U_μ. The following lemma can be considered as a first step in that direction. In this lemma which can be easily proved we will use the notation $E^\infty := \bigcap_{k\in\mathbb{N}} E^k$, with E^k being the class of all functions $v\in E$ having derivatives $\partial_t^\mu v \in E$ of order $\mu \le k$.

LEMMA (4.A). (i) Let h be any function defined on $\bar{\Omega}$ and with values in $\mathbb{C}$. Then the problem (P_0) admits the unique solution

$$U_0(x,t) = e^{-kt}h(x) \qquad (x\in\bar{\Omega},\ t \ge 0). \tag{4.4}$$

For each $x\in\bar{\Omega}$, and for any E in the collection $\mathcal{E}$, we have that $U_0(x,\cdot) \in E^\infty$, and hence that $U_0(x,\cdot) \in (L_+^1 \cap C_+^0)^\infty$.

(ii) The problems (P_μ) $(\mu = 1,\ldots,m)$ are of the particular form

$$\left.\begin{aligned} (k+\partial_t)U_\mu(x,t) &= f_\mu(x,t) && (x\in\Omega,\ t>o) \\ U_\mu(x,0) &= 0 && (x\in\Omega). \end{aligned}\right\} \tag{P}$$

Thus, if there exists a space $E\in\mathcal{E}$ such that $f_\mu(x,\cdot) \in E$ for each $x \in \bar{\Omega}$, then the unique solution of the problem (P) is given by

$$U_\mu(x,t) = (Gf_\mu)(x,t) = \int_0^t e^{-k(t-s)} f_\mu(x,s)ds \qquad (x \in \bar{\Omega},\ t \ge 0). \tag{4.5}$$

Consequently, $U_\mu(x,t) \in E_0^1$ for each $x \in \bar{\Omega}$.

A comparison of (P) with the original problems (P_μ) shows that

$$f_\mu(x,t) = \delta_{1\mu}\, r(x,t) - [K_c\Delta + K_1(k+\partial_t)]U_{\mu-1}(x,t) \qquad (\mu = 1,\dots,m). \quad (4.6)$$

Therefore, using an induction argument, we are able to express f_μ in terms of the known solution $U_0(x,t)$. To give an example, we shall compute $f_1(x,t)$. Recall the notations $g(t) = e^{-kt}$ $(t \geq o)$, and $g(t) = o$ $(t < o)$. Assuming now that $h \in C^2(\overline{\Omega})$, we deduce from (P_0) without any difficulties: $f_1(x,t) = r(x,t) - (K_c g)(t)(\Delta h)(x)$. Hence, it is immediately obvious that

$$U_1(x,t) = (Gr)(x,t) - (GK_c g)(t)(\Delta h)(x) \qquad (x\in\overline{\Omega},\ t \geq o), \qquad (4.7)$$

where $r(x,t)$ is required to satisfy the condition $r(x,\cdot) \in E$ $(x\in\overline{\Omega})$, for an arbitrary fixed $E\in\mathcal{E}$. In general, we obtain for $x\in\overline{\Omega}$ and $t \geq o$:

$$U_\mu(x,t) = (-1)^{\mu-1} \sum_{\nu=o}^{\mu-1} B_{\mu\nu}\, [(\Delta^\nu r)(x,t) - (K_c g)(t)(\Delta^{\nu+1}h)(x)], \qquad (4.8)$$

where we assume that $h \in C^{2\mu}(\overline{\Omega})$, and that $r \in C^{2\mu-2}(\overline{\Omega};E)$. Here $B_{\mu\nu}$ denotes a continuous linear operator depending on G, K_c and K_1, which will be specified in the following lemma.

LEMMA (4.B). Let E be any space of the collection $\mathcal{E}$. Let h belong to $C^{2m}(\overline{\Omega})$, and let r be an element of $C^{2m-2}(\overline{\Omega};E)$. Then we have

(i) in formula (4.8), each $B_{\mu\nu}$ $(\mu = 1,\dots,m;\ \nu = o,\dots,\mu-1)$ is a continuous linear operator, mapping from E into E^1_o. Its definition is given recurringly by the following rule

$$\left.\begin{aligned}
&B_{10} = G \quad \text{and} \quad B_{\mu+1,o} = GK_1G^{-1}B_{\mu o} && (\mu = 1,\ldots,m-1),\\
&B_{\mu+1,\mu} = GK_cB_{\mu,\mu-1} && (\mu = 1,\ldots,m-1),\\
&B_{\mu+1,\nu} = GK_cB_{\mu,\nu-1} + GK_1G^{-1}B_{\mu\nu} && (\mu = 2,\ldots,m-1;\ \nu = 1,\ldots,\mu-1).
\end{aligned}\right\}(4.9)$$

(ii) $||B_{\mu\nu}|| \leq \binom{\mu-1}{\nu} ||G||^{\nu+1}\, ||K_c||^{\nu}\, ||K_1||^{\mu-\nu-1} \quad (\mu = 1,\ldots,m;\ \nu = o,\ldots,\mu-1).$

Proof. (i) Let us use an induction argument for the proof of the relations (4.9). To begin with, we let $\mu = 1$. Then it follows from (4.7 and 4.8) that

$$U_1(x,t) = (Gr)(x,t) - (GK_cg)(t)(\Delta h)(x) \overset{!}{=} B_{10}[r(x,t)-(K_cg)(t)(\Delta h)(x)],$$

and hence we have $B_{10} = G\colon E \to E_o^1$ in accordance with the assertion. Next, we let $1<\mu<m$ be arbitrary but fixed. By hypothesis, we may assume that $U_\mu(x,\cdot) \in E_o^1$ for each $x\in\bar{\Omega}$. Since this implies the relation $(k+\partial_t)U_\mu = G^{-1}U_\mu$, we deduce from (4.5 and 4.6) that

$$U_{\mu+1}(x,t) = -\,(GK_c\Delta U_\mu + GK_1G^{-1}U_\mu)(x,t).$$

Using now the proposed formula (4.8) for U_μ, we thus obtain

$$\begin{aligned}
U_{\mu+1}(x,t) = (-1)^{\mu-1} \sum_{\nu=o}^{\mu-1} [&(GK_cB_{\mu\nu}\Delta^{\nu+1}r)(x,t) + (GK_1G^{-1}B_{\mu\nu}\Delta^{\nu}r)(x,t)\\
&- (GK_cB_{\mu\nu}K_cg)(t)(\Delta^{\nu+2}h)(x) - (GK_1G^{-1}B_{\mu\nu}K_cg)(t)(\Delta^{\nu+1}h)(x)].
\end{aligned}$$

After some rearrangements involving elementary substitutions in the summation variable, we see that this can be written in the equivalent form

$$(-1)^{\mu} \sum_{\nu=1}^{\mu-1} [(GK_c B_{\mu,\nu-1} + GK_1 G^{-1} B_{\mu\nu})(\Delta^{\nu} r)(x,t)$$

$$- (GK_c B_{\mu,\nu-1} K_c + GK_1 G^{-1} B_{\mu\nu} K_c) g(t) \cdot (\Delta^{\nu+1} h)(x)]$$

$$+ (-1)^{\mu} [(GK_c B_{\mu,\mu-1} \Delta^{\mu} r)(x,t) + (GK_1 G^{-1} B_{\mu o} r)(x,t)]$$

$$+ (-1)^{\mu} [(GK_c B_{\mu,\mu-1} K_c g)(t)(\Delta^{\mu+1} h)(x) + GK_1 G^{-1} B_{\mu o} K_c g)(t)(\Delta h)(x)].$$

Replacing now in formula (4.8) the index μ by $\mu+1$, and comparing the result with the above relation, we obtain the desired formulae (4.9). This completes the proof of the assertion (i).

(ii) In order to prove the proposed estimate, we shall use once more an induction argument. Starting with $\mu=1$, it follows from (4.9) that $||B_{10}|| = ||G||$ which is in accordance with the assertion. Assume now that $\nu=o$ and $\mu>1$ are given. Since every operator $B_{\mu\nu}$ is headed by the term G, we deduce that

$$||B_{\mu+1,o}|| \le ||G|| \, ||K_1|| \, ||G^{-1} B_{\mu o}|| \le ||G|| \, ||K_1|| \binom{\mu-1}{o} ||K_1||^{\mu-1} = \binom{\mu}{o} ||G|| \, ||K_1||^{\mu}$$

which is also in accordance with the assertion. Here we have used the induction hypothesis for $||B_{\mu o}||$. Next, we consider the case, where $\nu = \mu-1$ ($\mu = 2,\dots,m$). Again, it follows from (4.9) and from the induction hypothesis that

$$||B_{\mu+1,\mu}|| \le ||G|| \, ||K_c|| \, ||B_{\mu,\mu-1}|| \le ||G||^{\mu+1} \, ||K_c||^{\mu},$$

which is exactly the assertion. Finally, if we assume that $1 \le \nu \le \mu-1$, we obtain

$$||B_{\mu+1,\nu}|| \le ||G|| \, ||K_c|| \, ||B_{\mu,\nu-1}|| + ||G|| \, ||K_1|| \, ||G^{-1}B_{\mu\nu}||$$

$$\le ||G|| \, ||K_c|| \binom{\mu-1}{\nu-1} ||G||^{\nu} ||K_c||^{\nu-1} \, ||K_1||^{\mu-\nu}$$

$$+ ||G|| \, ||K_1|| \binom{\mu-1}{\nu} ||G||^{\nu} ||K_c||^{\nu} ||K_1||^{\mu-\nu-1}$$

$$= \binom{\mu}{\nu} ||G||^{\nu+1} \, ||K_c||^{\nu} \, ||K_1||^{\mu-\nu} ,$$

which is again in accordance with the assertion. Thus the proof of Lemma (4.B) is complete.

We conclude this section with the remark that in view of (4.8) and (4.9) the solutions U_μ of (P_μ) $(\mu = o,\ldots,m)$ are completely determined. Since $m \ge o$ was arbitrarily chosen, this procedure can be continued - at least theoretically - in order to obtain all higher order terms of the outer expansion.

4.2. THE INNER EXPANSION

In contrast to the notation used in Section 4.1, we now let Ω denote a bounded open <u>interval</u>, which will be considered as a subset of the Euclidean space $\mathbb{R}$. Without any loss of generality, we assume that $\Omega = (o,1)$. Otherwise we may apply a homothetic transformation to Ω. Further notations are accepted from the preceding section.

Recalling that, in general, the outer expansion (4.1) does not satisfy the prescribed boundary conditions at $x=o$ and $x=1$, namely

$$U(j,t;\varepsilon) = g_j(t) \quad (t \ge o; \quad j = o,1), \qquad (4.10)$$

we are forced to introduce boundary correction terms, by which a satisfactory asymptotic approximation of the solution u will be induced, not only outside some boundary layer region but also on the boundary lines itselves.

Generally speaking, there exists no global algorithm for the determination of the proper boundary correction terms, although some general outlines concerning this point already have been developed. We may refer to the work of W. Eckhaus [15], for instance. However, it seems to be more or less a question of experiences, how to find the proper approach fitting to the specific problem.

As for the problem (P_ε) under consideration, it is reasonable to consider a partition of the solution u into four terms, namely

$$u(x,t;\varepsilon) = U(x,t;\varepsilon) + V^o(x,t;\varepsilon) + V^1(x,t;\varepsilon) + Z(x,t;\varepsilon). \tag{4.11}$$

Here $U(x,t;\varepsilon)$ denotes the outer expansion which was determined in the previous section. The boundary correction terms $V^j(x,t;\varepsilon)$ are required to force the boundary conditions $g_j(t)$ on the boundary lines $x = j$ $(j = o,1)$. The remainder function $Z(x,t;\varepsilon)$ gives a measure for the magnitude of the difference between the exact solution $u(x,t;\varepsilon)$, and the approximate solution

$$\bar{u}(x,t;\varepsilon) = U(x,t;\varepsilon) + V^o(x,t;\varepsilon) + V^1(x,t;\varepsilon). \tag{4.12}$$

Clearly, if ε^m is the highest power involved in the outer expansion, then $Z(x,t;\varepsilon)$ should be expected to be bounded of order $O(\varepsilon^{m+1})$ as $\varepsilon \to o+$, possibly uniformly with respect to x and t.

In order to get conditions for the determination of V^j and Z, we substitute (4.11) in $M_\varepsilon[u] = \varepsilon r(x,t)$, and observe that U is completely fixed by (P_μ) $(\mu = o,\ldots,m)$. This yields:

$$\left.\begin{aligned} 0 = \varepsilon^{m+1} & [K_c \partial_x^2 U_m + K_1 G^{-1} U_m](x,t) \\ & + (k+\partial_t)V^o(x,t;\varepsilon) + \varepsilon[K_c \partial_x^2 V^o + K_1(k+\partial_t)V^o](x,t;\varepsilon) \\ & + (k+\partial_t)V^1(x,t;\varepsilon) + \varepsilon[K_c \partial_x^2 V^1 + K_1(k+\partial_t)V^1](x,t;\varepsilon) \\ & + M_\varepsilon[Z](x,t;\varepsilon) \qquad (o < x < 1;\ t > o). \end{aligned}\right\} \tag{4.13}$$

Now, by following the outlines of the matching principle (see W. Eckhaus [15]), we introduce into our calculus new (stretching or rapidly varying) variables

$$\tilde{x}_o := x/\sqrt{\varepsilon}, \quad \tilde{x}_1 := (1-x)/\sqrt{\varepsilon} \quad (x \in [o,1], \quad \varepsilon > o), \tag{4.14}$$

to which it may be referred as to the <u>inner</u> <u>variables</u> (in contrast to the outer variable x). Substituting $\tilde{x}_j$ in V^j, for simplicity, we shall continue to denote V^j by the same symbol: $V^j = V^j(\tilde{x}_j,t;\varepsilon)$ $(j = o,1)$. Furthermore, corresponding to the outer expansion (4.1), we are going to look for an <u>inner</u> <u>expansion</u>

$$V^j(\tilde{x}_j,t;\varepsilon) = \sum_{\mu=o}^{m} \varepsilon^\mu V^j_\mu(\tilde{x}_j,t) \quad (j = o,1). \tag{4.15}$$

Its necessity may be motivated with the intention to determine boundary correction terms which have, in case of $\varepsilon \to o+$, the same order of magnitude as the outer expansion has. This is yielded by a substitution of (4.15) in (4.13) and a comparison of equal powers of ε and of same indexes j.

Having carried out the details, we will obtain the following abstract parabolic differential equations:

$$\left.\begin{aligned}
&(k + \partial_t + K_c \partial^2_{\tilde{x}_j}) V^j_o (\tilde{x}_j, t) = o && (\tilde{x}_j > o,\ t > o) \\
& && j = o, 1 \\
&(k + \partial_t + K_c \partial^2_{\tilde{x}_j}) V^j_\mu (\tilde{x}_j, t) = -K_1 (k + \partial_t) V^j_{\mu-1} (\tilde{x}_j, t) && (\tilde{x}_j > o,\ t > o),
\end{aligned}\right\} \quad (4.16)$$

$$\left.\begin{aligned}
M_\varepsilon[Z](x,t;\varepsilon) = -\varepsilon^{m+1} \{ &(K_c \partial^2_x + K_1 G^{-1}) U_m(x,t) \\
&+ K_1 (k+\partial_t) [V^o_m(x/\sqrt{\varepsilon}, t) + V^1_m((1-x)/\sqrt{\varepsilon}, t)] \} \\
&(o < x < 1, \quad t > o).
\end{aligned}\right\} \quad (4.17)$$

Summarizing to this point, we thus have established differential conditions for the determination of V^j and Z, where we did not yet take care of the prescribed initial and boundary data. To this last purpose, we would like to remark that by (P_ε) and by (4.11) we are caused to require the following conditions

$$\begin{aligned}
h(x) &= \sum_{\mu=o}^{m} \varepsilon^\mu [U_\mu(x,o) + V^o_\mu(\tilde{x}_o, o) + V^1_\mu(\tilde{x}_1, o)] + Z(x,o;\varepsilon) \quad (o<x<1), \\
g_o(t) &= \sum_{\mu=o}^{m} \varepsilon^\mu [U_\mu(o,t) + V^o_\mu(o,t) + V^1_\mu(1/\sqrt{\varepsilon}, t)] + Z(o,t;\varepsilon) \quad (t \ge o), \\
g_1(t) &= \sum_{\mu=o}^{m} \varepsilon^\mu [U_\mu(1,t) + V^o_\mu(1/\sqrt{\varepsilon}, t) + V^1_\mu(o,t)] + Z(1,t;\varepsilon) \quad (t \ge o).
\end{aligned}$$

(4.18)

Again, we may satisfy these conditions by a termwise comparison of same powers of ε. Combining now the resulting relations with the differential equations (4.16) and (4.17), we are led to a complete set of initial boundary value problems for the determination of V^j_μ $(\mu = o,\ldots,m;\ \ j = o,1)$, and Z. In the following theorem we shall give the corresponding formulations.

THEOREM (4.C). Let the outer expansion

$$U(x,t;\varepsilon) = \sum_{\mu=o}^{m} \varepsilon^{\mu} U_{\mu}(x,t) \qquad (o \le x \le 1, \quad t \ge o)$$

be uniquely determined by the solutions U_{μ} of (P_{μ}) $(\mu = o,\dots,m)$. Then the inner expansion of a solution u to the problem (P_{ε}) is obtained from setting

$$u(x,t;\varepsilon) = U(x,t;\varepsilon) + V^{o}(x,t;\varepsilon) + V^{1}(x,t;\varepsilon) + Z(x,t;\varepsilon),$$

and by using the method of stretched variables $\tilde{x}_o := x/\sqrt{\varepsilon}$ and $\tilde{x}_1 := (1-x)/\sqrt{\varepsilon}$ $(o \le x \le 1)$. In particular, assuming that

$$V^{j}(x,t;\varepsilon) = \sum_{\mu=o}^{m} \varepsilon^{\mu} V^{j}_{\mu}(\tilde{x}_j,t) \qquad (o \le x \le 1, \quad t \ge o),$$

we are led to the following initial boundary value problems for the determination of V^j and Z:

$$\left.\begin{aligned} &(k + \partial_t + K_c \partial^2_{\tilde{x}_j})\, V^{j}_{o}(\tilde{x}_j,t) = o && (\tilde{x}_j > o,\ t>o),\\ &V^{j}_{o}(o,t) = g_j(t) - h(j)e^{-kt} && (t \ge o),\\ &V^{j}_{o}(\tilde{x}_j,o) = o && (\tilde{x}_j > o), \end{aligned}\right\} \qquad (Q^{j}_{o})$$

$$\left.\begin{aligned}
&(k + \partial_t + K_c \partial^2_{\tilde{x}_j})\, V^j_\mu(\tilde{x}_j,t) = -K_1(k+\partial_t)V^j_{\mu-1}(\tilde{x}_j,t) && (\tilde{x}_j > o,\ t>o),\\
&V^j_\mu(o,t) = -U_\mu(j,t) && (t \geq o),\\
&V^j_\mu(\tilde{x}_j,o) = o && (\tilde{x}_j > o),
\end{aligned}\right\} \quad (Q^j_\mu)$$

where $\mu = 1,\ldots,m$ and $j = o,1$. Furthermore,

$$\left.\begin{aligned}
&M_\varepsilon[Z](x,t;\varepsilon) = -\varepsilon^{m+1} w(x,t;\varepsilon) && (o<x<1,\ t>o),\\
&Z(o,t;\varepsilon) = -\sum_{\mu=o}^{m} \varepsilon^\mu\, V^1_\mu(1/\sqrt{\varepsilon},t) && (t \geq o),\\
&Z(1,t;\varepsilon) = -\sum_{\mu=o}^{m} \varepsilon^\mu\, V^o_\mu(1/\sqrt{\varepsilon},t) && (t \geq o),\\
&Z(x,o;\varepsilon) = o && (o<x<1),
\end{aligned}\right\} \quad (Z)$$

where the right-hand side function w is defined by

$$\begin{aligned}
w(x,t;\varepsilon) = {} & (K_c\partial^2_x + K_1G^{-1})U_m(x,t)\\
& + K_1(k+\partial_t)\,[V^o_m(x/\sqrt{\varepsilon},t) + V^1_m((1-x)/\sqrt{\varepsilon},t)].
\end{aligned} \qquad (4.19)$$

In principle, the problem (Z) is an abstract two-point boundary value problem of the type (3.12) which was extensively studied in Sections 3.4 and 3.5. We shall come back to this problem in chapter 6.

As for the problems (Q^j_μ) $(\mu = o,\ldots,m)$, we would like to remark that they are non-well posed, because we are free to impose an additional boundary condition onto each of them. It is here that one needs the so called Matching condition which will compensate the missing boundary condition. To be more precise, we shall add to each problem (Q^j_μ) the following

(MC) MATCHING CONDITION: $\lim_{y\to\infty} V_\mu^j(y,t) = o \quad (t \geq o).$

From this condition, the boundary layer terms V_μ^j are forced to an exponential rate of decay when y tends to infinity. More loosely speaking, the Matching condition keeps the influence of the boundary correction term V_μ^j on the opposite part of the boundary as small as possible.

Considering now the problems $(Q_\mu^j + MC)$ $(\mu = o,\dots,m)$, we are completely in the situation described by the abstract two-point boundary value problem (3.11). A particular investigation of this relationship shall be executed in the next chapter.

REMARKS. 1°. We would like to point out that the underlying ideas of our approach can also be applied to various related problems. Take the example that the given data $h(x)$ and $g_j(t)$ depend analytically upon the parameter ε. Then, in formulae (4.18), we may replace h and g_j by the respective Taylor expansions about $\varepsilon=o$:

$$h(x;\varepsilon) = \sum_{\mu=o}^{m} \varepsilon^\mu h_\mu(x) + \varepsilon^{m+1} \tilde{h}_{\mu+1},$$

$$g_j(t;\varepsilon) = \sum_{\mu=o}^{m} \varepsilon^\mu g_{j\mu}(t) + \varepsilon^{m+1} \tilde{g}_{j,\mu+1}.$$

Equating now same powers of ε, the functions h_μ and $g_{j\mu}$ will appear in the initial and boundary conditions of (Q_μ^j). A similar consideration also is valid for the right-hand side function $r(x,t)$.

2°. Another admissible - though more difficult - generalization is the consideration of an arbitrary bounded open domain $\Omega \subset \mathbb{R}^n$ having a smooth boundary. In this example, one can introduce boundary correction terms via local cards in nearly the same way as we did. In the usual parametrization of the boundary manifold $\partial\Omega$ the local coordinates are lying in an (n-1)-dimensional hyperplane. Clearly, in this case the stretched coordinate, perpendicular to that hyperplane, is the proper inner variable. For more details, we refer the reader to A. van Harten [25].

BIBLIOGRAPHICAL NOTES

Sec. 4.1 and 4.2. The motion of a viscous fluid of small viscosity was the first area of research leading to singular perturbation problems. In particular, L. Prandtl's boundary layer theory [41] had considerable influence on the development of singular perturbation methods. Once developed, the methods have been applied to problems in other fields.

The use of outer and inner expansions is known from the early beginning of singular perturbation theory, and also the investigation of the boundary layer by means of a stretching transformation has been practised since that time.

One of the most important techniques in singular perturbations, namely the matching of the inner and outer expansions, was first proposed by L. Prandtl [41]. It was developed by subsequent investigators, in particular S. Kaplun and P.A. Lagerstrom [29,30], and most recently investigated in considerable detail by L.E. Fraenkel [18] and in less detail but greater generality by W. Eckhaus [15].

Singular perturbations in connection with two-point boundary value problems for ordinary linear differential equations have been successfully discussed in the pioneering paper of M.I. Višik and L.A. Lyusternik [48]. The introduction of the two different stretching transformations (4.14) has been performed more or less in the spirit of their approach. The more related problems for ordinary linear integro-differential equations and for parabolic partial differential equations have been studied by N.P. Vekua [47] and by F. Hoppenstaedt [27] respectively. For further references and applications of singular perturbation methods to problems in different fields, the reader is also referred to the surveying article by A. Erdélyi [17].

5 The structure of the boundary layer

In this chapter we shall apply the existence theory for abstract two-point boundary value problems to the study of the boundary layer problems (Q_μ^j), to which we will add the Matching condition (MC). To begin with, we shall individually treat each of the problems (Q_0^j) and (Q_1^j), and then we will turn over to the general case, which will be easily handled by generalizing the foregoing investigations.

In the first section, the treatment of the problems (Q_0^j) is rather simple because the corresponding differential equations are associated only with homogeneous right-hand sides. Here we shall establish a general existence and uniqueness result for solutions to (Q_0^j). We shall also state the Compatibility conditions (H3) to which the initial and boundary data have to be exposed.

The second section is concerned with regularity properties of the solutions to (Q_0^j). In particular, a Hölder condition shall be proved which will constitute the core of the further existence theory.

In Section 5.3, an existence and uniqueness result for solutions to the problems (Q_1^j) will be established. Again, the smoothness of these solutions plays a fundamental role, and is basic for the treatment of the general case.

In Section 5.4, the arguments and results of the preceding part are extended to the general problems (Q_μ^j) $(\mu \geq 2)$. Its solutions will complete the determination of the boundary layer.

5.1. AN EXISTENCE AND UNIQUENESS RESULT

In this section we shall discuss existence and uniqueness of solutions to the boundary layer problems (Q_o^j) $(j = o,1)$, to which we will add the Matching condition (MC). For simplicity we shall denote by y each of the two inner variables $\tilde{x}_o = x/\sqrt{\varepsilon}$, and $\tilde{x}_1 = (1-x)/\sqrt{\varepsilon}$ simultaneously. Observing these notations, we are concerned with the following initial boundary value problems:

$$\left.\begin{array}{ll} (k+\partial_t + K_c\partial_y^2)V_o^j(y,t) = o & (y > o,\ t > o), \\ V_o^j(o,t) = g_j(t) - h(j)e^{-kt} & (t \geq o), \\ V_o^j(y,o) = o & (y > o), \\ \lim\limits_{y\to\infty} V_o^j(y,t) = o & (t \geq o). \end{array}\right\} (Q_o^j + MC)$$

The basic situation of this section is assumed to be the following. Firstly, the space $E\in\mathcal{E}$ will be arbitrary but fixed. The operators G, $K_c = cI + K_o$, and $A = GK_c$ are required to satisfy the Sector condition (H1). Notations of function spaces, and of the operator family $S_o(y)$ $(y>o)$ will be the same as in Section 3.1.

Secondly, the initial and boundary data are required to satisfy the following

(H3) COMPATIBILITY CONDITIONS. Let g_j $(j = o,1)$ belong to E^1, and assume that $h \in C[o,1]$ is given such that $g_j(o) = h(j)$ $(j = o,1)$ holds.

These conditions assure the existence and uniqueness of solutions to the problems under consideration, as we will see in the following main theorem.

THEOREM (5.A). Let the Sector condition (H1) and the Compatibility conditions (H3) be satisfied, and suppose that the Root definition (H2) holds. Then the initial and boundary value problems $(Q_o^j + MC)$ $(j = o,1)$ have unique solutions $V_o^j \in C^2(\mathbb{R}_+;E) \cap C_o(\overline{\mathbb{R}_+};E)$ satisfying $V_o^j(y,\cdot) \in E_o^1$ for every $y \geq o$. These solutions admit a representation in the form

$$\begin{aligned} V_o^j(y,t) &= S_o(y)[g_j(t) - h(j)e^{-kt}] \\ &= (2\pi i)^{-1} \int_L e^{y\sqrt{-\lambda}} R'(\lambda;A)[g_j - U_o(j,\cdot)](t) \frac{d\lambda}{\lambda} . \end{aligned} \tag{5.1}$$

In addition, we have $V_o^j \in C^\infty(\mathbb{R}_+;E) \cap L^p(\mathbb{R}_+;E)$ for each $1 \leq p \leq \infty$.

Proof. We shall divide the proof into two parts, the existence part, and the uniqueness part.

(a) Existence. We set $v_j(t) := g_j(t) - h(j)e^{-kt}$ $(t \geq o;\ j = o,1)$. In virtue of the Compatibility conditions (H3), we see that $v_j \in E_o^1$. Hence, we conclude from Theorem (3.A.iii) that the functions $S_o(y)v_j$ are right-continuous at $y=o$, having a function limit $S_o(o+)v_j = v_j \in E_o^1$. This together with (3.9), and with Theorem (3.A.i) implies that

$$S_o(y)v_j \in C^\infty(\mathbb{R}_+;E) \cap C_o(\overline{\mathbb{R}_+};E) \cap L^p(\mathbb{R}_+;E) \quad (1 \leq p \leq \infty),$$

and that also $S_o(y)v_j \in E_o^1$ for every $y \geq o$. Thus we obtain $S_o(y)v_j|_{t=o} = o$ for every $y \geq o$. Recalling that $G^{-1}: E_o^1 \to E$ is a bijection and applying Theorem (3.A.ii), we get $G^{-1}S_o(y)v_j + K_c\partial_y^2 S_o(y)v_j = o$ $(y > o)$. Hence, we have proved that $V_o^j(y,t) := (S_o(y)v_j)(t)$ $(j = o,1)$ solves the problems $(Q_o^j + MC)$.

(b) Uniqueness. Suppose that there exists another solution of $(Q_o^j + MC)$, say $W_o^j(\cdot): \overline{\mathbb{R}_+} \to E_o^1$, with the property $W_o^j \in C^2(\mathbb{R}_+;E) \cap C_o(\overline{\mathbb{R}_+};E)$. Clearly, W_o^j satisfies the equation $W_o^j(y) + A\partial_y^2 W_o^j(y) = o \quad (y > o)$. But in virtue of Lemma (3.G), this equation is equivalent to

$$[I - (-A)^{1/2} \partial_y] [I + (-A)^{1/2} \partial_y] W_o^j(y) = o \quad (y > o). \tag{5.2}$$

Here we have used the fact that the continuous linear operator $(-A)^{1/2}$ commutes with ∂_y. Using now standard arguments from spectral theory together with the Matching condition (MC), we see that there is only one possibility in order to satisfy the relation (5.2), namely

$$W_o^j(y) - (-A)^{1/2} \partial_y W_o^j(y) = o \quad (y > o). \tag{5.3}$$

To continue the uniqueness proof, we shall consider the function $T^j(y,z) := S_o(y-z)W_o^j(z) \quad (o \le z \le y < \infty)$. Differentiating with respect to z, and observing (5.3) yields

$$\partial_z T^j(y,z) = - S_o'(y-z)(-A)^{1/2}\partial_z W_o^j(z) + S_o(y-z)\partial_z W_o^j(z) \quad (o<z<y). \tag{5.4}$$

Since $(-A)^{1/2} = S_1(o)$ (see (3.15)), we deduce in much the same manner as in the proof of Theorem (3.H.i) that $S_o'(y-z)\, S_1(o) = S_o(y-z)$. This, combined with (5.3) gives $\partial_z T^j(y,z) = o \quad (o<z<y)$. Integrating over (o,y) it results that $T^j(y,z) = T^j(y,o) \quad (o \le z \le y)$. In particular, letting $z = y$ and observing that $W_o^j(o) = V_o^j(o)$, we obtain

$$\begin{aligned} T^j(y,y) &= S_o(o)\, W_o^j(y) = W_o^j(y) = T^j(y,o) \\ &= S_o(y)\, W_o^j(o) = S_o(y)\, V_o^j(o) = V_o^j(y) \quad (y \ge o). \end{aligned}$$

This completes the proof of Theorem (5.A).

REMARKS. 1°. We would like to point out that in $(Q_o^j + MC)$, each relation has to be understood in the sense of a pointwise equality with respect to every $y \geq o$ and every $t \geq o$. This is even true in case of $E = L_+^p$ $(1 \leq p \leq \infty)$.

2°. In view of Corollary (3.E) we may obtain an existence and uniqueness result corresponding to that of Theorem (5.A), if in (H3) we cancel the conditions $g_j(o) = k(j)$ $(j = o,1)$. However, in this case the boundary conditions fail to be satisfied for $t = o$.

5.2. THE HÖLDER REGULARITY OF V_o^j

In this section we shall collect some preliminary results which will be needed for the consideration of the initial boundary value problems $(Q_1^j + MC)$. As we have already mentioned, these problems are abstract two-point boundary value problems of the type (3.11). The right-hand side functions have to be specified in the following way:

$$f_1^j(y)(t) := - K_c^{-1} K_1 (k+\partial_t) V_o^j(y,t) \quad (j = o,1) \tag{5.5}$$

Here, and in the sequel, we agree with the notations used in the preceding section.

In view of the existence results proved in Theorem (3.M), we should much like to know whether the Hölder condition (F) of Section 3.5 could be verified for f_1^j, or not. The answer to this question will be given in the following theorem which is the basic result of the present section.

THEOREM (5.B). Let the Sector condition (H1) and the Compatibility conditions (H3) be satisfied, and suppose that the Root definition (H2) holds. Then, for every $o < \alpha \leq 1$ there exists a constant $C > o$ such that

$$||f_1^j(y) - f_1^j(z)|| \leq C H_\alpha(\eta) \, |y-z|^\alpha \quad \text{for all } y,z > o,$$

where η is any positive number satisfying $o < \eta \leq \min\{y,z\}$, and where $H_\alpha(\eta)$ is defined by

$$H_\alpha(\eta) := e^{-\eta\sqrt{\rho}\,\sin\theta/2} + \eta^{-\alpha} e^{-\eta\sqrt{\rho}/2\,\sin\theta/2} \quad (o < \eta < \infty).$$

Clearly, the constant C does not depend upon y,z and η.

The proof of this theorem is more involved, and therefore, we shall divide it into several parts. We shall start with a rather technical result on Hölder continuity.

PROPOSITION (5.C). Let L be the oriented path of Fig. 1, and suppose that the Root definition (H2) holds. Let $\lambda \in L$ be represented in the form $\lambda = re^{i\psi}$ $(\rho \leq r < \infty, \ \theta \leq \psi \leq 2\pi - \theta)$. Then, for every $o < \alpha \leq 1$ there exists a constant $C > o$ not depending upon λ and η such that $|1 - e^{\eta\sqrt{-\lambda}}| \leq C\eta^\alpha r^{\alpha/2}$ for all $\eta > o$ and any $\lambda \in L$.

Proof. Using the representation (3.3) for the square root $\sqrt{-\lambda}$, it is immediately obvious that for all $\lambda \in L$

$$|1-e^{\eta\sqrt{-\lambda}}| = [1-2e^{-\eta\sqrt{r}\,\sin\psi/2}\cos(\eta\sqrt{r}\cos\psi/2) + e^{-2\eta\sqrt{r}\,\sin\psi/2}]^{1/2}.$$

Letting $\zeta := \eta\sqrt{r}$, we may define

$$w(\zeta) := [1-2e^{-\zeta \sin \psi/2} \cos(\zeta \cos \psi/2) + e^{-2\zeta \sin \psi/2}]/\zeta^{2\alpha} \quad (\zeta > o).$$

Clearly, the mapping $w(\cdot): \mathbb{R}_+ \to \overline{\mathbb{R}_+}$ is continuous, and it satisfies $\lim_{\zeta\to\infty} w(\zeta) = o$. Consequently, on each interval $[\zeta_o,\infty)$ $(\zeta_o > o)$, it can be bounded uniformly with respect to ζ, and also uniformly with respect to $\psi \in [\theta, 2\pi - \theta]$. Thus, in order to complete the proof of the proposition, it suffices to show that the limit $\lim_{\zeta\to o+} w(\zeta)$ also exists. But this can be done by an application of the L'Hospital rule, from which it follows that

$$\lim_{\zeta\to o+} w(\zeta) = \begin{cases} o & \text{for each } o<\alpha<1, \\ 1 & \text{for } \alpha = 1, \\ \infty & \text{for } \alpha > 1. \end{cases}$$

This completes the proof of Proposition (5.C).

Next, we would like to have a more convenient representation formula for the functions $f_1^j(y)$. To this purpose, we should write the relations (5.1) in the equivalent form $V_o^j(y,t) = A\, S_o(y)A^{-1}\, [g_j(t) - h(j)e^{-kt}]$ $(y > o)$. Using the definition (3.1), and setting $g(t) := e^{-kt}$ $(t \geq o)$ respectively $g(t) = o$ $(t < o)$, this yields

$$f_1^j(y) = -(2\pi i)^{-1}\, K_c^{-1}\, K_1 K_c \int_L e^{y\sqrt{-\lambda}}\, R'(\lambda;A)A^{-1}[g_j-h(j)g]\lambda^{-1}d\lambda. \qquad (5.6)$$

We are now in a position to perform the <u>proof of Theorem</u> (5.B). To begin with, we may assume without any loss of generality that $y>o$ and $z>o$ satisfy the relation $y \geq z$. Since K_c^{-1}, K_1, K_c, $R'(\lambda;A)$, and $A^{-1}[g_j-h(j)g]$ are bounded in their respective norms, we can find a constant

$C>o$ not depending upon y and z, such that

$$\|f_1^j(y) - f_1^j(z)\| \le C \int_L |1 - e^{(y-z)\sqrt{-\lambda}}| \, e^{-z\sqrt{r}\sin\theta/2} \, r^{-1} \, |d\lambda|.$$

Here we have used the representation formula (5.6), as well as the substitution $\lambda = re^{i\psi}$ $(\rho \le r < \infty;\ \theta \le \psi \le 2\pi - \theta)$.

Next, dividing L into the arcs Γ_ρ and $\Gamma_\pm$, and using the results of Proposition (5.C), we continue to estimate the above inequality by

$$C|y-z|^\alpha (e^{-z\sqrt{\rho}\sin\theta/2} + \int_\rho^\infty e^{-z\sqrt{r}\sin\theta/2} \, r^{-(1-\alpha/2)} \, dr). \tag{5.7}$$

Finally, observing that the last integral can be bounded by

$$2e^{-z\sqrt{\rho}/2\ \sin\theta/2} \int_0^\infty e^{-z\tau/2\ \sin\theta/2} \, \tau^{\alpha-1} \, d\tau = C(\alpha) z^{-\alpha} \, e^{-z\sqrt{\rho}/2\ \sin\theta/2}$$

$$(o < \alpha \le 1), \tag{5.8}$$

where $C(\alpha)$ is a suitable constant, we thus have proved the assertion of Theorem (5.B).

REMARK. Reconsidering the estimate (5.7), we can also prove that

$$2\int_\rho^\infty e^{-z\sqrt{r}\sin\theta/2} \, r^{-(1-\alpha/2)} \, dr \leqq 2e^{-\beta z} \int_0^\infty e^{-z\tau\,(\sin\theta/2 - \beta/\sqrt{\rho})} \, \tau^{\alpha-1} d\tau$$

$$\le C(\alpha,\beta) z^{-\alpha} \, e^{-\beta z},$$

where β is any constant satisfying $o < \beta < \sqrt{\rho}\sin\theta/2$. In view of this fact, we may replace in Theorem (5.B) the Hölder coefficient $H_\alpha(\eta)$ by

$$H_{\alpha\beta}(\eta) := e^{-\eta\sqrt{\rho}\,\sin\theta/2} + \eta^{-\alpha}e^{-\beta\eta} \quad (o<\eta<\infty;\quad o<\beta<\sqrt{\rho}\,\sin\theta/2). \tag{5.9}$$

5.3. THE BOUNDARY LAYER TERMS V_1^j

In this section, we shall continue to determine the general structure of the boundary layer $V^j(x,t;\varepsilon)$ $(j = o,1)$. More precisely, we shall establish an existence and uniqueness result for the particular terms $V_1^j(\tilde{x}_j,t)$. The next step, leading from V_1^j to the determination of V_μ^j for an arbitrary $\mu>1$, will be done in Section 5.4.

In what follows, we agree with the notations used in Section 5.1. In particular, y denotes again either of the two variables $\tilde{x}_o$ and $\tilde{x}_1$. With this notation, we are concerned with the following initial boundary value problem.

$$\left.\begin{array}{ll} (k + \partial_t + K_c\partial_y^2)\, V_1^j(y,t) = (K_c f_1^j(y))(t) & (y>o,\ t>o), \\ V_1^j(o,t) = -\,U_1(j,t) & (t \geq o), \\ V_1^j(y,o) = o & (y > o), \\ \lim\limits_{y\to\infty} V_1^j(y,t) = o & (t \geq o). \end{array}\right\} \quad (Q_1^j + MC)$$

Here f_1^j $(j = o,1)$ are the functions defined by (5.5) (or equivalently by (5.6)), and U_1 is given by (4.8).

Concerning the differentiability properties of U_μ $(\mu \geq 1)$, we will make a convenient regularity assumption below.

$(H)_\mu$ REGULARITY CONDITION. Let r belong to $C^{2\mu-2}([o,1];E)$, and suppose that $h \in C^{2\mu}[o,1]$.

For convenience, we have used here the notation $C^o([o,1];E)$ for the class $C([o,1];E)$.

One of the main results of the present section is the following theorem.

THEOREM (5.D). Let the Sector condition (H1) and the Compatibility conditions (H3) be satisfied, and suppose that the Root definition (H2) and the Regularity condition $(H)_1$ hold. Then the initial boundary value problems $(Q_1^j + MC)$ $(j = o,1)$ have unique solutions $V_1^j \in C^2(\mathbb{R}_+;E) \cap C_o(\overline{\mathbb{R}_+};E)$ satisfying $V_1^j(y,\cdot) \in E_o^1$ for every $y \geq o$. These solutions admit a representation in the form

$$V_1^j(y,t) = (R_1(y)f_1^j)(t) - S_o(y)U_1(j,t) \quad (y \geq o,\ t \geq o), \tag{5.10}$$

where f_1^j and U_1 are respectively determined by (5.6) and (4.8). In addition, we have $S_o(\cdot)U_1(j,t) \in C^\infty(\mathbb{R}_+;E) \cap L^p(\mathbb{R}_+;E)$ $(1 \leq p \leq \infty)$.

Proof. Let us begin with the existence proof. To this purpose we let

$$V_1^j(y,t) := W_1^j(y,t) + W_2^j(y,t) \tag{5.11}$$

where W_1^j is assumed to be the solution of the homogeneous problem

$$\left.\begin{aligned} &(k + \partial_t + K_c\partial_y^2)\, W_1^j(y,t) = o && (y > o,\ t > o),\\ &W_1^j(o,t) = -\,U_1(j,t) && (t \geq o),\\ &W_1^j(y,o) = o && (y > o),\\ &\lim_{y\to\infty} W_1^j(y,t) = o && (t \geq o), \end{aligned}\right\} \tag{5.12}$$

and where W_2^j should be considered as the solution of the inhomogeneous problem

$$\left.\begin{aligned}
&(k + \partial_t + K_c \partial_y^2)\, W_2^j(y,t) = (K_c f_1^j(y))(t) && (y > o,\ t > o),\\
&W_2^j(o,t) = o && (t \geq o),\\
&W_2^j(y,o) = o && (y > o),\\
&\lim_{y\to\infty} W_2^j(y,t) = o && (t \geq o).
\end{aligned}\right\} \tag{5.13}$$

Clearly, thanks to the linearity of the decomposition (5.11), a combination of (5.12) and (5.13) leads back to the problems $(Q_1^j + MC)$. Now, from Lemma (4.B) and from the assumption $(H)_1$, we get $U_1(j,\cdot) \in E_o^1$ $(j = o,1)$. Hence, we may apply Theorem (5.A) in order to see that the problems (5.12) are uniquely solved by $W_1^j(y,t) = - S_o(y)U_1(j,t)$ $(y \geq o,\ t \geq o)$. In addition, we obtain the regularity properties

$$W_1^j \in C^\infty(\mathbb{R}_+;E) \cap L^p(\mathbb{R}_+;E) \cap C_o(\overline{\mathbb{R}_+};E) \quad (1 \leq p \leq \infty).$$

As for the solutions of the problems (5.13), we would like to note that in virtue of Theorem (5.B) each f_1^j $(j = o,1)$ satisfies the Hölder condition (F) of Section 3.5. Furthermore, from Theorem (3.A.i), and from the representation (5.6) we may deduce that

$$f_1^j \in L^p(\mathbb{R}_+;E) \cap C_o(\mathbb{R}_+;E) \quad (1 \leq p < \infty). \tag{5.14}$$

Letting now $W_2^j(y,t) := (R_1(y)f_1^j)(t)$ $(y \geq o,\ t \geq o)$, it follows by Theorem (3.M) that $W_2^j \in C^2(\mathbb{R}_+;E) \cap C(\overline{\mathbb{R}_+};E)$, and $W_2^j(y,\cdot) \in E_o^1$ $(y > o)$. Thus, we may apply the operator G^{-1} to W_2^j which together with Theorem (3.M.ii) yields

$$\left.\begin{aligned} &(G^{-1} + K_c\partial_y^2)W_2^j(y,t) = (K_c\ f_1^j(y))(t) && (y>o,\ t>o) \\ &W_2^j(y,o) = o && (y>o). \end{aligned}\right\} \qquad (5.15)$$

Finally, combining (5.14) with Theorem (3.K.i), we conclude that $W_2^j(o,t) = o = \lim_{y\to\infty} W_2^j(y,t)$ $(t \ge o)$. This shows that W_2^j is a solution to (5.13), and hence the existence proof is complete.

In order to prove uniqueness, we may assume that $\tilde{V}_1^j \in C^2(\mathbb{R}_+;E) \cap C_o(\overline{\mathbb{R}_+};E)$ is another solution of $(Q_1^j + MC)$ satisfying $\tilde{V}_1^j(y,\cdot) \in E_o^1$ $(y \ge o)$. Letting $W^j(y,t) := V_1^j(y,t) - \tilde{V}_1^j(y,t)$, it is immediately obvious that W^j is a solution of (5.12), where it is assumed that $U_1 = o$. Applying now Theorem (5.A), we see that $W^j \doteq o$, and this completes the proof of Theorem (5.D).

COROLLARY (5.E). Under the assumptions of Theorem (5.D), we also have that $V_1^j \in L^p(\mathbb{R}_+;E)$ $(1 \le p \le \infty)$.

<u>Proof.</u> Clearly, by Theorem (5.D) it suffices to prove that $W_2^j = R_1(\cdot)f_1^j$ belongs to $L^1(\mathbb{R}_+;E)$, because this already implies that

$$W_2^j \in L^1(\mathbb{R}_+;E) \cap C_o(\overline{\mathbb{R}_+};E) \subset \bigcap_{p\ge 1} L^p(\mathbb{R}_+;E).$$

Now, since W_2^j belongs to $C_o(\overline{\mathbb{R}_+};E)$, it follows from Pettis' theorem that the mapping $W_2^j\colon \overline{\mathbb{R}_+} \to E$ is strongly measurable. Hence it remains to show that the integral $\int_0^\infty ||W_2^j(y,\cdot)||dy$ is finite. Recalling the definition of $R_1(y)$, we see that

$$||W_2^j(y,\cdot)|| \le \frac{1}{2}\{\int_0^\infty ||S_1(z)||\ ||f_1^j(y+z)||dz + \int_0^y ||S_1(y-z)||\ ||f_1^j(z)||dz\}$$
$$+ \frac{1}{2}\int_0^\infty ||S_1(y+z)||\ ||f_1^j(z)||dz.$$

Observing that $S_1 \in L^1(\mathbb{R}_+;B(E))$, and $f_1^j \in L^1(\mathbb{R}_+;E)$, the desired result is now immediately obtained by an application of Fubini's theorem. This completes the proof.

Reviewing the proof outlines of Theorem (5.D), we will see that the Hölder regularity of the functions f_1^j is crucial for the existence part. Apparently, for further investigations on the boundary layer terms V_μ^j $(\mu \ge 2)$, one needs a similar regularity result for the right-hand side functions

$$f_2^j(y)(t) := -\,K_c^{-1}\ K_1(k+\partial_t)\ V_1^j(y,t) \quad (j = o,1). \tag{5.16}$$

Unfortunately, this time a corresponding proof is more involved, because the boundary layer functions V_1^j consist not only of the rather simple terms $W_1^j(y,t) = -\,S_o(y)U_1(j,t)$ but also of the much more complicated terms $W_2^j(y,t) = (R_1(y)f_1^j)(t)$. However, we shall prove a regularity result which corresponds to that of Theorem (5.B).

To begin with, we will bring f_2^j into a more convenient form. As for the first term of V_1^j, namely W_1^j, we shall adapt the representation formula (5.6). For the second term W_2^j, we may use a combination of (5.15) and (3.34). Putting this together, we will obtain

$$\left.\begin{aligned} f_2^j(y) &= K_c^{-1}K_1K_c\ \{S_o(y)\ A^{-1}U_1(j,\cdot) - f_1^j(y) \\ &+ \frac{1}{2}\int_y^\infty S_o'(z-y)[f_1^j(z)-f_1^j(y)]dz + \frac{1}{2}\int_o^y S_o'(y-z)[f_1^j(z)-f_1^j(y)]dz \\ &- \frac{1}{2}\int_o^\infty S_o'(z+y)[f_1^j(z)+f_1^j(y)]dz\}. \end{aligned}\right\} \quad (5.17)$$

Using this representation, we can establish the basic regularity result as follows.

THEOREM (5.F). Let the Sector condition (H1) and the Compatibility conditions (H3) be satisfied, and suppose that the Root definition (H2) and the Regularity condition $(H)_1$ hold. Then we have

(i) $f_2^j \in C_o(\mathbb{R}_+;E) \cap L^p(\mathbb{R}_+;E)$ for any $1 \le p < \infty$. In addition, for every $\alpha \in (o,1)$, the functions $y^\alpha f_2^j(y)$ are bounded in norm uniformly with respect to $y \in \overline{\mathbb{R}_+}$.

(ii) For every $\alpha \in (o,1)$, there exists a constant $C > o$ such that

$$||f_2^j(y) - f_2^j(y')|| \le C\ H_{\alpha\beta}(\eta)\,|y-y'|^\alpha \quad \text{for all } y,y' > o,$$

where η is any positive number satisfying $o < \eta \le \min\{y,y'\}$, and where $H_{\alpha\beta}(\eta)$ is defined by

$$H_{\alpha\beta}(\eta) := e^{-\beta\eta} + \eta^{-\tilde{\alpha}}e^{-\beta\eta/2} \quad (o < \eta < \infty), \tag{5.18}$$

with arbitrary constants $\tilde{\alpha}$ and β satisfying $\alpha < \tilde{\alpha} < 1$ and $o < \beta < \sqrt{\rho}\,\sin\frac{\theta}{2}$.

In order to prove this theorem, we shall proceed in several steps. For the sake of simplicity, let us introduce the following notations:

$$I_1(y) := \int_y^\infty S_o'(z)\,[f_1^j(z+y) - f_1^j(z-y)]\,dz \qquad (y > o),$$

$$I_2(y) := \int_o^y S_o'(z)\,[f_1^j(z+y) - f_1^j(y)]\,dz \qquad (y > o),$$

$$I_3(y) := \int_o^y S_o'(z)\,[f_1^j(y-z) - f_1^j(y)]\,dz \qquad (y > o),$$

$$I_4(y) := 2\,S_o(y)\,f_1^j(y) \qquad (y > o).$$

Now, it is easily to see that

$$\left.\begin{aligned} f_2^j(y) &= K_c^{-1}K_1K_c\,\{S_o(y)A^{-1}U_1(j,\cdot) - f_1^j(y) \\ &\quad + \tfrac{1}{2}\,I_1(y) + \tfrac{1}{2}\,I_2(y) + \tfrac{1}{2}\,I_3(y) + \tfrac{1}{2}\,I_4(y)\}. \end{aligned}\right\} \quad (5.19)$$

Now we shall consider each term in formula (5.19) individually. Starting with $I_1(y)$, we shall prove:

PROPOSITION (5.G). Under the assumptions of Theorem (5.F), for every $\alpha \in (o,1)$, there exists a constant $C > o$ such that

(i) $||I_1(y)|| \le C\,e^{-y\sqrt{\rho}\,\sin\theta/2}\,(y^{1-\alpha} + y^{-\alpha}) \quad (o < y < \infty)$,

(ii) $||I_1(y) - I_1(y')|| \le C\,H_{\alpha\beta}(\eta)\,|y-y'|^\alpha$ for all $y,y' > o$, where $H_{\alpha\beta}(\eta)$ is defined by (5.18).

Proof. (i) Let $\alpha \in (o,1)$ be arbitrary but fixed. Using the definitions (3.4 and 3.5), we see that $S_o'(z) = I_\rho(z;1) + I_+(z;1) + I\_(z;1) \quad (o<z<\infty)$. From this we obtain

$$||S_0'(z)|| \le C\, e^{-z\sqrt{\rho}\,\sin\theta/2}\,(1 + z^{-1}) \quad (o < z < \infty). \tag{5.20}$$

Setting $\alpha' := 1 - \alpha \in (o,1)$ and combining the estimate (5.20) with the result of Theorem (5.B) one finds:

$$\left.\begin{aligned} &||S_0'(z)\,[f_1^j(z+y) - f_1^j(z-y)]|| \\ &\le C\, y^{\alpha'}\, e^{-z\sqrt{\rho}\,\sin\theta/2}(1+z^{-1})\,[e^{-(z-y)\sqrt{\rho}\,\sin\theta/2} \\ &+ (z-y)^{-\alpha'}\, e^{-(z-y)\sqrt{\rho}/2\,\sin\theta/2}] \quad (o < y < z < \infty). \end{aligned}\right\} \tag{5.21}$$

Integrating now over (y,∞), it follows that $||I_1(y)|| \le C\, e^{-y\sqrt{\rho}\,\sin\theta/2}\,(y^{\alpha'} + y^{\alpha'-1}) \quad (o < y < \infty)$, as claimed in assertion (i).

(ii) Again, we let $\alpha \in (o,1)$ be arbitrary but fixed. Given any $y,y' > o$, we shall assume without loss of generality that $y \ge y'$. Applying the triangle inequality, we easily deduce that

$$\left.\begin{aligned} ||I_1(y) - I_1(y')|| \le &\int_y^\infty ||S_0'(z)||\,||f_1^j(z+y) - f_1^j(z+y')||dz \\ &+ \int_y^\infty ||S_0'(z)||\,||f_1^j(z-y) - f_1^j(z-y')||dz \\ &+ \int_{y'}^y ||S_0'(z)||\,||f_1^j(z+y') - f_1^j(z-y')||dz. \end{aligned}\right\}$$

Essentially, by following the outlines of the preceding proof part, we will see that the first two integrals to the right of the above inequality can be bounded by

$$C|y-y'|^{\alpha}\,[e^{-y'\sqrt{\rho}\,\sin\theta/2} + (y')^{-\alpha}\, e^{-y'\sqrt{\rho}/2\,\sin\theta/2}].$$

Therefore, we have only to deal with the last integral. Setting $\alpha' := 1 - \alpha \in (o,1)$, and using an estimate corresponding to (5.21), we infer that

$$\left.\begin{aligned} &\int_{y'}^{y} ||S_o'(z)||\, ||f_1^j(z+y') - f_1^j(z-y')||dz \\ &\leq C(y')^{\alpha'} e^{-y'\sqrt{\rho}\sin\theta/2} \{\int_o^{y-y'} e^{-z\sqrt{\rho}\sin\theta/2} dz \\ &+ \int_o^{y-y'} (z+y')^{-1} e^{-z\sqrt{\rho}\sin\theta/2} dz + \int_o^{y-y'} z^{-\alpha'} e^{-z\sqrt{\rho}/2\sin\theta/2} dz \\ &+ \int_o^{y-y'} z^{-\alpha'}(z+y')^{-1} e^{-z\sqrt{\rho}/2\sin\theta/2} dz\}. \end{aligned}\right\} (5.22)$$

Denoting the integrals within the brackets respectively by J_μ $(\mu = 1,\ldots,4)$, we conclude that

$$J_1 \leq (|y-y'|^\alpha\sqrt{\rho}\ \sin\theta/2)^{-1}(1-e^{-(y-y')\sqrt{\rho}\sin\theta/2})|y-y'|^\alpha \leq C|y-y'|^\alpha,$$

$$J_2 \leq C(y')^{-1}\ |y-y'|^\alpha,$$

$$J_3 \leq \int_o^{y-y'} z^{-\alpha'} dz = (1-\alpha')^{-1}\ (y-y')^{1-\alpha'},$$

$$J_4 \leq (y')^{-1} \int_o^{y-y'} z^{-\alpha'} dz = (1-\alpha')^{-1}(y')^{-1}(y-y')^{1-\alpha'}.$$

Putting these estimates into (5.22), and observing that $(y')^{\alpha'} e^{-y'\sqrt{\rho}\sin\theta/2} \leq C\, e^{-\beta y'}$, where C is a suitable constant, and where β is limited by $o < \beta < \sqrt{\rho}\sin\theta/2$, we have

$$\int_y^{y'} ||S_o'(z)||\, ||f_1^j(z+y') - f_1^j(z-y')||dz \leq C|y-y'|^\alpha[e^{-\beta y'} + (y')^{-\alpha}e^{-\beta y'/2}].$$

This completes the proof of Proposition (5.G).

To continue the considerations of formula (5.19), we now shall turn to the integrals $I_2(y)$, $I_3(y)$, $I_4(y)$. Repeating the ideas and arguments of the proof to Proposition (5.G), we shall obtain the following result.

COROLLARY (5.H). Under the assumptions of Theorem (5.F), for every $\alpha \in (0,1)$, there exists a constant $C > o$ such that for any $o < y < \infty$:

(i) $||I_2(y)|| \le C\, e^{-y\sqrt{\rho}/2 \sin\theta/2}\,(1+y^\alpha)$,

$||I_3(y)|| \le C\, e^{-y\sqrt{\rho}/2 \sin\theta/2}\,(y^{1+\alpha} + y^\alpha + y^{1-\alpha} + y^{-\alpha})$,

$||I_4(y)|| \le C\, e^{-\beta y}\,(1 + y^{-\alpha})$,

(ii) $||I_\mu(y) - I_\mu(y')|| \le C\, H_{\alpha\beta}(\eta)\,|y-y'|^\alpha$ for all $y,y' > o$ and $\mu = 2,3,4$,

where $H_{\alpha\beta}(\eta)$ is defined by (5.18).

Proceeding with the functions $W^j(y,t) := S_o(y)\, A^{-1}\, U_1(j,t)$ in formula (5.19), we may apply Theorem (3.A) in order to see that $W^j \in L^p(\mathbb{R}_+;E) \cap C_o(\mathbb{R}_+;E)$ $(1 \le p < \infty)$. Furthermore, using the estimates (5.7 and 5.8), we deduce that

$$||W^j(y,\cdot)|| \le C(e^{-y\sqrt{\rho}\sin\theta/2} + y^{-\alpha}\, e^{-y\sqrt{\rho}/2 \sin\theta/2}) \qquad (y > o),$$

where $\alpha \in (0,1)$ can be arbitrarily chosen. This implies $\sup_{y\in\mathbb{R}_+} ||y^\alpha\, W^j(y,\cdot)|| < \infty$ $(o < \alpha < 1)$.

Next, we observe that by (5.6) the relations $f_1^j(y) = -\,K_c^{-1} K_1 K_c S_o(y) A^{-1}[g_j - h(j)g]$ hold. Clearly, by repeating the arguments used for W^j, we easily get

$$f_1^j(\cdot) \in L^p(\mathbb{R}_+;E) \cap C_o(\mathbb{R}_+;E) \quad (1\le p<\infty),\ \sup_{y\in\mathbb{R}_+} ||y^\alpha\, f_1^j(y)|| < \infty \quad (o<\alpha<1).$$

Recalling now the results of Theorem (5.B), we conclude the proof of Theorem (5.F) by collecting the details proved above, and putting them into formula (5.19).

5.4. THE GENERAL CASE

In the present section we shall be concerned with the determination of the general boundary layer terms v_μ^j. More precisely, we shall consider the initial boundary value problems

$$\left.\begin{aligned}
&(k + \partial_t + K_c \partial_y^2)\, v_\mu^j(y,t) = (K_c f_\mu^j(y))(t) && (y > o,\ t > o),\\
&v_\mu^j(o,t) = -\, U_\mu(j,t) && (t \geq o),\\
&v_\mu^j(y,o) = o && (y > o),\\
&\lim_{y\to\infty} v_\mu^j(y,t) = o && (t \geq o),
\end{aligned}\right\} (Q_\mu^j + MC)$$

where $1 \leq \mu \leq m$ is arbitrary but fixed, and where we agree with the notations used in the preceding sections of this chapter.

To be consequent, we would like to begin with the case where $\mu=2$. Thus, let us assume that the Regularity condition $(H)_2$ holds (which is defined in Section 5.3). Consequently, the boundary data $U_2(j,t)$ $(j = o,1)$ satisfy the particular conditions $U_2(j,\cdot) \in E_o^1$. This is easily obtained by applying Lemma (4.B) to the representation formula (4.8).

Next, we recall the definitions of the right-hand side functions f_2^j given in (5.17). Comparing the results of Theorem (5.B) on f_1^j, and the results of Theorem (5.F) on f_2^j, we claim that f_1^j and f_2^j are equipped with essentially the same regularity properties.

Summarizing to this point, we may infer that the circumstances of the problems $(Q_2^j + MC)$ are exactly comparable with the circumstances of the problems $(Q_1^j + MC)$. Hence, we may apply Theorem (5.D) to $(Q_2^j + MC)$ in order to see that there exist unique solutions

$$v_2^j(y,t) = (R_1(y)f_2^j)(t) - S_o(y)U_2(j,t) \quad (y \geq o,\ t \geq o),$$

satisfying $v_2^j \in C^2(\mathbb{R}_+;E) \cap C_o(\overline{\mathbb{R}_+};E) \cap L^p(\mathbb{R}_+;E)$ $(1 \leq p \leq \infty)$, and $v_2^j(y,\cdot) \in E_o^1$ $(y \geq o)$.

Moreover, corresponding to formulae (5.16 and 5.17), we may define right-hand side functions f_3^j $(j = o,1)$ by means of the known functions f_2^j, namely

$$\left.\begin{aligned} f_3^j(y) &:= - K_c^{-1}K_1(k+\partial_t)\, v_2^j(y,\cdot) \\ &= K_c^{-1}K_1K_c\{S_o(y)\, A^{-1}U_2(j,\cdot) - f_2^j(y) \\ &+ \frac{1}{2}\int_y^\infty S_o'(z-y)[f_2^j(z)-f_2^j(y)]dz + \frac{1}{2}\int_o^y S_o'(y-z)[f_2^j(z)-f_2^j(y)]dz \\ &- \frac{1}{2}\int_o^\infty S_o'(z+y)[f_2^j(z) + f_2^j(y)]dz\}. \end{aligned}\right\}$$

Clearly, Theorem (5.F) also applies to f_3^j, and hence we are completely able to proceed to the consideration of the succeeding initial boundary value problem $(Q_3^j + MC)$.

Thus, by further continuation of this induction principle, we will obtain a general existence and uniqueness result which shall be stated in the following theorem.

THEOREM (5.I). Let the Sector condition (H1) and the Compatibility conditions (H3) be satisfied, and suppose that the Root definition (H2) and the Regularity condition $(H)_\mu$ $(\mu \geq 1)$ hold. Then, the initial boundary value problems $(Q_\mu^j + MC)$ $(j = 0,1)$ have unique solutions $V_\mu^j \in C^2(\mathbb{R}_+;E) \cap C_o(\overline{\mathbb{R}_+};E) \cap L^p(\mathbb{R}_+;E)$ $(1 \leq p \leq \infty)$ satisfying $V_\mu^j(y,\cdot) \in E_o^1$ for every $y \geq o$. These solutions admit a representation in the form

$$V_\mu^j(y,t) = (R_1(y)f_\mu^1)(t) - S_o(y)U_\mu(j,t) \quad (y \geq o,\ t \geq o), \tag{5.23}$$

where U_μ is determined by (4.8), and where f_μ^j is recurringly defined by the rule

$$\left.\begin{aligned} f_\mu^j(y) &= - K_c^{-1}K_1(k+\partial_t)\, V_{\mu-1}^j(y,\cdot) \\ &= K_c^{-1}K_1K_c\{S_o(y)\, A^{-1}\, U_{\mu-1}(j,\cdot) - f_{\mu-1}^j(y) \\ &+ \frac{1}{2}\int_y^\infty S_o'(z-y)\,[f_{\mu-1}^j(z) - f_{\mu-1}^j(y)]dz \\ &+ \frac{1}{2}\int_o^y S_o'(y-z)\,[f_{\mu-1}^j(z) - f_{\mu-1}^j(y)]dz \\ &- \frac{1}{2}\int_o^\infty S_o'(y+z)\,[f_{\mu-1}^j(z) + f_{\mu-1}^j(y)]dz\}. \quad (y > o;\ \mu \geq 2). \end{aligned}\right\} \tag{5.24}$$

The particular functions f_1^j are given by formula (5.6). In addition, we have $S_o(\cdot)\ U_\mu(j,t) \in C^\infty(\mathbb{R}_+;E)$.

This theorem generalizes the results of Theorem (5.D) and of Corollary (5.E) in an obvious way, and there is no further proof needed. Equivalently, the results of Theorem (5.F) generalize to those of the following corollary.

COROLLARY (5.K). Under the assumptions of Theorem (5.I), we have

(i) $f_\mu^j \in C_o(\mathbb{R}_+;E) \cap L^p(\mathbb{R}_+;E)$ for any $1 \leq p < \infty$. In addition, for every $\alpha \in (o,1)$, the functions $y^\alpha f_\mu^j(y)$ are bounded in norm uniformly with respect to $y \in \overline{\mathbb{R}_+}$.

(ii) For every $\alpha \in (o,1)$, there exists a constant $C > o$ such that $||f_\mu^j(y) - f_\mu^j(y')|| \leq C\, H_{\alpha\beta}(\eta)\, |y-y'|^\alpha$ for all $y,y' > o$, where η is any positive number satisfying $o < \eta \leq \min\{y,y'\}$, and where $H_{\alpha\beta}(\eta)$ is defined by (5.18).

BIBLIOGRAPHICAL NOTES

Sec. 5.1. The results of this section are due to the author [23]. In that paper, the singular perturbation problem (P_ε) was solved by a zeroth order approximate solution. However, the proof of the uniqueness part in Theorem (5.A) differs from the earlier one. Here we follow the outlines of a standard procedure which is used for uniqueness proofs in the theory of linear equations of evolution (see for instance T. Kato [31]).

Sec. 5.2. Hölder continuity of solutions to inhomogeneous linear evolution equations was first established by T. Kato [32] for the case that these solutions are associated with holomorphic semigroups (see also T. Kato [31] and R.W. Carroll [8]). Theorem (5.B) is closely related to results of this nature in the sense that uniform Hölder continuity can only be obtained on a semi-infinite interval $[\eta,+\infty)$ $(\eta>o)$ which is bounded away from zero. However, the growth property of the Hölder constant seems to be new within this context.

Sec. 5.3 and 5.4. The results of these sections generalize those of the author's paper [23], where the zeroth order initial boundary value problems $(Q_o^j + MC)$ were only solved. In particular, the remarkable results of Theorem (5.F) and Corollary (5.K) regarding the boundedness of the functions

$f_{\mu}^{j}(y)$ in a right neighbourhood of $y = o$, lead to a new proof for the uniform boundedness of the remainder function under much weaker assumptions than in our earlier paper [23].

6 The remainder problem

Summarizing the results of chapters 4 and 5, we have constructed a function $\bar{u} = \bar{u}(x,t;\varepsilon)$ which is an approximate solution to the singular perturbation problem (P_ε) (at least asymptotically for $\varepsilon \to o+$). As we have seen, this function $\bar{u}$ admits a representation in the form (4.12). Clearly, one is mainly interested in the question of the degree to which a genuine solution $u(x,t;\varepsilon)$ to (P_ε) is approximated by $\bar{u}$. As a measure of the quality of approximation, we shall consider the natural distance function in the Banach space $C([o,1];E)$. Here the space E is fixed within the collection $\mathcal{E}$ by the regularity properties of the given data.

Assuming that u is a genuine solution to (P_ε), we may define

$$Z(x,t;\varepsilon) := u(x,t;\varepsilon) - \bar{u}(x,t;\varepsilon). \qquad (*)$$

Recall that a substitution of (*) into problem (P_ε) leads to the remainder problem (Z) which was formulated in Theorem (4.C). In the present chapter, we shall treat this problem.

In the first section, we shall establish further results on abstract two-point boundary value problems which are closely related to earlier results on this matter, obtained in Sections 3.4 and 3.5. The main idea is now to transform the remainder problem (Z) into a problem of type (3.12), where the operator A has to be replaced by another appropriate operator $\tilde{A}$. To assure boundedness of $\tilde{A}$, we have to require further regularity properties for the convolution operator K_1.

Using the results of the first section, in the second section we shall establish a general existence and uniqueness result for solutions to the remainder problem (Z). Taking into account that the approximate solution $\bar{u}(x,t;\varepsilon)$ is already uniquely determined, the relation (*) thus defines a unique solution to the singular perturbation problem (P_ε). In our approach, solutions are only obtained for sufficiently small ε lying in a right neighborhood of the origin. Bounds for this neighborhood will be given in terms of the operators K_c, K_1 and G.

Finally, Section 6.3 will be concerned with the question of the quality of the approximate solution $\bar{u}$. As already mentioned, we shall measure the degree of approximation in terms of the norm $\sup_{0 \leq x \leq 1} ||Z(x,\cdot;\varepsilon)||_E$. This norm will be seen to the bounded of order ε^{m+1} as $\varepsilon \to o+$, where the natural number m is determined by the maximal power of ε appearing in the outer and inner expansions.

6.1. FURTHER INVESTIGATIONS OF TWO-POINT BOUNDARY VALUE PROBLEMS

This section is concerned with two-point boundary value problems of the form

$$\left.\begin{aligned} &A\partial_y^2 u(y) + (I + \varepsilon G K_1 G^{-1}) u(y) = A f(y) \quad (o < y < \delta), \\ &u(o) = o = u(\delta), \end{aligned}\right\} \qquad (6.1)$$

on a Banach space E of the collection $\mathcal{E}$. Here G, $K_c = cI + K_o$, K_1, and $A = GK_c$ are the continuous linear operators in E which were introduced in Sections 2.3 and 4.1, while $f(y)$ is a prescribed function, defined on (o,δ) and with values in E. Furthermore, we shall assume that $\varepsilon_o > o$ is a fixed number whereas ε can be arbitrarily chosen with $o < \varepsilon \leqq \varepsilon_o$. The parameter δ should be considered as a function of ε, $\delta = \delta(\varepsilon)$, which is required to be nonnegative and bounded from below by $\delta_0 > 0$, uniformly with respect to $0 \le \varepsilon \le \varepsilon_0$. In particular, we think of the function

$$\delta(\varepsilon) = 1/\sqrt{\varepsilon} \quad (0 < \varepsilon \le \varepsilon_0) \tag{6.2}$$

as a realization of δ. Restrictions on ε_0 will be imposed below.

The main purpose of this section is to provide sufficient conditions on ε_0 and K_1 such that the closed linear operator $I + \varepsilon G K_1 G^{-1}: E_0^1 \to E$ admits on E a bounded inverse. If this is the case, we may multiply the differential equation in (6.1) from the left by $(I + \varepsilon G K_1 G^{-1})^{-1}$. Obviously, the resulting problem can be considered as a two-point boundary value problem of the type (3.12), where the continuous linear operator A has to be replaced by the continuous linear operator $\tilde{A} := (I + \varepsilon G K_1 G^{-1})^{-1} A$.

In the next step, we will demonstrate that the Sector condition (H1) guarantees the sector S_θ to be contained in the resolvent set of $\tilde{A}$, and the reduced resolvent function $R'(\lambda;\tilde{A})$ to be bounded uniformly with respect to $\lambda \in S_\theta$. By the way of replacing in the definitions of $S_\mu(y)$ $(\mu = 0,1,2)$ and of $R_2(y)$, the operator A by $\tilde{A}$, we then may apply the whole theory of the foregoing chapters to the problem (6.1). In particular, we shall obtain from the Theorems (3.L) and (3.N) a general existence and uniqueness result for the solutions to (6.1).

We shall start the study of the operator $I + \varepsilon G K_1 G^{-1}$ with the proof of the following important commutation property.

LEMMA (6.A). Let the kernel function k_1 belong to $L^1(\mathbb{R}) \cap L^\infty_{loc}(\mathbb{R}_+)$. Then we have $K_1 G^{-1} v = (k + \partial_t)\tilde{K}_1 v$ for every $v \in E_0^1$, where the continuous linear operator $\tilde{K}_1: E \to E$ is defined by

$$(\tilde{K}_1 v)(t) := \int_{-\infty}^{t} k_1(s)v(t-s)ds \quad (t \geq o).$$

Proof. First of all, it is easily checked that $\tilde{K}_1$ is a continuous linear operator mapping from E into E. Next, we shall show that $\partial_t(\tilde{K}_1 v)(t) = (\tilde{K}_1 \partial_t v)(t)$. From this the assertion of the lemma is immediately obtained. Hence, let $v \in E_o^1$ be arbitrary but fixed, and let $h \neq o$ be any real number. We consider the difference quotient

$$h^{-1}[(\tilde{K}_1 v)(t+h) - (\tilde{K}_1 v)(t)]$$

$$= h^{-1}\{\int_{-\infty}^{t} k_1(s)[v(t+h-s)-v(t-s)]ds + \int_{t}^{t+h} k_1(s)v(t+h-s)ds\} \quad (t \geq o).$$

Since the limit $\lim_{h\to o} h^{-1}[v(t+h)-v(t)] = (\partial_t v)(t)$ exists in E, we may use a standard device of integration theory in order to see that the first integral in the above relation tends to $(\tilde{K}_1 \partial_t v)(t)$ as h tends to zero. Next, the continuity of v together with the property $v(o) = o$ imply that

$$|v(t+h-s)| = |v(t+h-s)-v(o)| \leq \eta \quad (|t+h-s| \leq K(\eta)),$$

where $\eta > o$ is any given number, and where $K(\eta)$ is a suitable constant depending on v and $t \geq o$. Letting $|h| \leq K(\eta)$, it turns out that

$$\limsup_{h\to o} |h^{-1} \int_{t}^{t+h} k_1(s)v(t+h-s)ds| \leq \eta \cdot \operatorname{ess.sup}_{s\in[t-K,t+K]} |k_1(s)|.$$

Since $\eta > o$ was arbitrary, this gives $\lim_{h\to o} h^{-1} \int_{t}^{t+h} k_1(s)v(t+h-s)ds = o$, which completes the proof.

REMARK. The condition on $k_1(t)$, namely $k_1 \in L^{\infty}_{loc}(\mathbb{R}_+)$, may be weakened to the condition: 'almost every point $t \in \overline{\mathbb{R}_+}$ is a Lebesgue point of $k_1(t)$'. In this case we obtain

$$\lim_{h \to o} h^{-1} \int_t^{t+h} |k_1(s)| ds = |k_1(t)| \qquad (\text{a.e. } t \in \overline{\mathbb{R}_+}),$$

and hence the proof of Lemma (6.A) still remains valid.

Continuing the investigations of the operator $I + \varepsilon G K_1 G^{-1}$, we shall introduce the notation E^* for the topological dual of E. Clearly, if $E = L^p_+$ $(1 \leq p < \infty)$, we have $E^* = L^q_+$ $(p^{-1} + q^{-1} = 1)$ which is contained in the collection $\mathcal{E}$. However, if $E \subseteq L^{\infty}_+$, then we have $L^1_+ \subsetneqq E^*$, and therefore it follows that $E^* \notin \mathcal{E}$. In view of this observation we may define

$$\check{E} := \begin{cases} E^* & \text{if } E^* \in \mathcal{E}, \\ L^1_+ & \text{if } E^* \notin \mathcal{E}. \end{cases}$$

Using this notation, we now shall prove an extension result for the closed linear operator GK_1G^{-1}.

LEMMA (6.B). Assume that the kernel function $\check{k}_1(s) := k_1(-s)$ belongs to $L^1(\mathbb{R}) \cap L^{\infty}_{loc}(-\infty, o) \cap \check{E}$, and define the operator B by

$$(Bv)(t) := (K_1 v)(t) - e^{-kt}(K_1 v)(o) \qquad (v \in E,\ t \geq o), \tag{6.3}$$

where $(K_1v)(o) := \int_o^\infty k_1(-s)v(s)ds$. Then B is a continuous extension of the closed linear operator GK_1G^{-1} satisfying $B_{|E_o^1} = GK_1G^{-1}$ and having a bound

$$||B||_{B(L_+^p)} \leq (||K_1|| + (pk)^{-1/p}\,||\check{k}_1||_q) \quad (1 \leq p \leq \infty,\ p^{-1} + q^{-1} = 1). \quad (6.4)$$

Proof. First we shall prove the estimate (6.4). Clearly, since $K_1 \in B(E)$, and since $e^{-kt} \in (L_+^1 \cap C_+^o) = E_\cap$, it is immediately obvious that $B \in B(E)$. Let us now consider the two cases according to $E = L_+^p$ $(1 \leq p < \infty)$, and to $E \subseteq L_+^\infty$. In the first case, we have for any $v \in L_+^1$:

$$\begin{aligned}||e^{-kt}(K_1v)(o)||_1 &= \int_o^\infty |e^{-kt}\int_o^\infty k_1(-s)v(s)ds|dt \\ &\leq k^{-1}||v||_1 \operatorname*{ess.sup}_{t\geq o} |\check{k}_1(t)|,\end{aligned}$$

respectively for any $v \in L_+^p$ $(1 < p < \infty)$:

$$\begin{aligned}||e^{-kt}(K_1v)(o)||_p &= [\int_o^\infty |e^{-kt}\int_o^\infty k_1(-s)v(s)ds|^p dt]^{1/p} \\ &\leq [\int_o^\infty e^{-pkt}dt\ (\int_o^\infty |k_1(-s)|^q ds)^{p/q}\int_o^\infty |v(s)|^p ds]^{1/p} \\ &\leq (pk)^{-1/p}||v||_p\,||\check{k}_1||_q.\end{aligned}$$

Using this together with the definition (6.3), we obtain the estimate (6.4). In the second case, we have for any $v \in E \subseteq L_+^\infty$:

$$||e^{-kt}\ (K_1v)(o)||_\infty = \operatorname*{ess.sup}_{t\geq o} |e^{-kt}\int_o^\infty k_1(-s)v(s)ds| \leq ||v||_\infty\,||\check{k}_1||_1.$$

Again, this leads to the estimate (6.4).

Next, we shall prove the property $B_{|E_o^1} = GK_1G^{-1}$. To this purpose, we assume that $v \in E_o^1$ is arbitrary but fixed. Then we may see by a substitution of variables that

$$(Bv)(t) = \int_{-\infty}^{t} k_1(s)v(t-s)ds - e^{-kt}\int_{-\infty}^{o} k_1(s)v(-s)ds$$

$$= (\tilde{K}_1 v)(t) - e^{-kt}(\tilde{K}_1 v)(o) \qquad (t \geq o).$$

From this relation it follows that $(Bv)(o) = o$. Moreover, by Lemma (6.A) we see that $\tilde{K}_1 v \in D(k+\partial_t) = E^1$. This yields $Bv \in E_o^1$, and hence $(Bv)(t) = (GG^{-1}Bv)(t) = G[(k+\partial_t)(\tilde{K}_1 v)(t) - (k+\partial_t)e^{-kt}(\tilde{K}_1 v)(o)] = (GK_1G^{-1}v)(t)$ $(t \geq o)$. Here we have used the result of Lemma (6.A), and the fact that $(k+\partial_t)e^{-kt} = o$. This completes the proof.

Observe that the boundedness of the operator B is a consequence purely of the condition $\check{k}_1 \in L^1(\mathbb{R}) \cap \check{E}$. Under this condition, using the fixed-point contraction theorem, we obtain the existence of the inverse operator $(I + \varepsilon B)^{-1}: E \to E$, for every ε satisfying $o \leq \varepsilon \leq \varepsilon_o < ||B||^{-1}$.

We shall use this observation for the proof of the following characterization of the reduced resolvent function

$$R'(\lambda;\tilde{A}) := [\lambda(I + \varepsilon B)^{-1} A - I]^{-1}. \tag{6.5}$$

LEMMA (6.C). Let the kernel function $\check{k}_1$ belong to $L^1(\mathbb{R}) \cap \check{E}$. Let $\lambda \in \mathbb{C}$ be an element of the resolvent set $\rho(A^{-1})$, and suppose that

$$o \leq \varepsilon \leq \varepsilon_o < \min \{||B||^{-1}, (||R'(\lambda;A)||\,||B||)^{-1}\}. \tag{6.6}$$

Then the reduced resolvent function (6.5) exists as a continuous linear operator on E, having a uniform bound for all $o \le \varepsilon \le \varepsilon_o$.

Proof. For an arbitrary but fixed $v \in E$, we may define $f \in E$ by $f := \lambda(I + \varepsilon B)^{-1}Av - v$. Now by hypothesis, the operators $I + \varepsilon B$ and $R'(\lambda;A)$ are homeomorphisms from E onto E, and hence we can write equivalently

$$v = R'(\lambda;A)(I + \varepsilon B)f + \varepsilon R'(\lambda;A)Bv. \tag{6.7}$$

Substituting $R'(\lambda;A)(I + \varepsilon B)f$ by $\tilde{f} \in E$, we may apply the method of successive approximations to (6.7). This yields

$$[\lambda(I + \varepsilon B)^{-1}A - I]^{-1}f = v = \sum_{\mu=o}^{\infty} [\varepsilon R'(\lambda;A)B]^{\mu} \tilde{f} \quad (o \le \varepsilon \le \varepsilon_o).$$

Thus, we have the existence of $R'(\lambda;\tilde{A}) \in B(E)$. Next, by (6.6)

$$||R'(\lambda;A)f|| \le \sum_{\mu=o}^{\infty} (\varepsilon_o ||R'(\lambda;A)|| \, ||B||)^{\mu} \, ||\tilde{f}|| \le \frac{||R'(\lambda;A)|| \, (1+\varepsilon_o ||B||)||f||}{1-\varepsilon_o \, ||R'(\lambda;A)|| \, ||B||},$$

which is valid uniformly with respect to $\varepsilon \in [o,\varepsilon_o]$. This concludes the proof.

Recalling now the uniform boundedness of $R'(\lambda;A)$ on the sector S_θ, we may ask, if there exists a sector $\tilde{S}_\theta \subset \mathbb{C}$ in the complex plane such that the reduced resolvent $R'(\lambda;\tilde{A})$ is also bounded uniformly with respect to $\lambda \in \tilde{S}_\theta$. We shall give an affirmative answer in the following corollary which sharpens the results of Lemma (6.C).

COROLLARY (6.D). Let the Sector condition (H1) be satisfied, and let the kernel function $\check{k}_1$ belong to $L^1(\mathbb{R}) \cap \check{E}$. Let $L \subset \mathbb{C}$ be given as in Fig. 1, and let $\Omega \subset \mathbb{C}$ denote the open domain surrounded by L and containing the origin. Let the positive constant C be defined by $C := \sup_{\lambda \in \overline{\Omega}} \|R'(\lambda;A)\|$, and assume that

$$0 < \varepsilon_0 < \min \{\|B\|^{-1}, (\|B\|C)^{-1}\}. \tag{6.8}$$

Then it follows that $\|R'(\lambda;\tilde{A})\| \leq C(1+\varepsilon_0\|B\|)/(1-\varepsilon_0\|B\|C)$, for all $\lambda \in \overline{\Omega}$, and all $0 \leq \varepsilon \leq \varepsilon_0$.

Proof. There is nothing to do, except to combine Corollary (2.N) with Lemma (6.C).

Next, we are interested in the relationship between the operators $I + \varepsilon GK_1G^{-1}$ and $(I + \varepsilon B)^{-1}$. On E_0^1, these operators are inverse to each other, which will be proved in the following lemma.

LEMMA (6.E). Let the kernel function $\check{k}_1$ belong to $L^1(\mathbb{R}) \cap L^\infty_{loc}(-\infty,0) \cap \check{E}$. Then for every $0 \leq \varepsilon \leq \varepsilon_0 < \|B\|^{-1}$, and for every $v \in E_0^1$ we have

(i) $(I + \varepsilon B)^{-1} (I + \varepsilon GK_1G^{-1})v = v$,

(ii) $(I + \varepsilon GK_1G^{-1})(I + \varepsilon B)^{-1}v = v$.

Proof. Let $v \in E_0^1$ be arbitrary but fixed. According to Lemma (6.B), we have the relation $GK_1G^{-1}v = Bv$, and hence the assertion (i) is proven.

Next, we shall show that $(I + \varepsilon B)^{-1}(E_o^1) \subset E_o^1$. Then, using Lemma (6.B) once again, we will obtain the assertion (ii). Hence, let us define

$$w := (I + \varepsilon B)^{-1} v \quad \text{and} \quad w_n := \sum_{\mu=o}^{n} (-1)^{\mu} (\varepsilon B)^{\mu} v \qquad (n \in \mathbb{N}).$$

Since $Bv = GK_1G^{-1}v \in E_o^1$, and since $o \leq \varepsilon \leq \varepsilon_o < ||B||^{-1}$, it follows that $w_n \in E_o^1$ $(n \in \mathbb{N})$ and $\lim_{n\to\infty} ||w-w_n|| = o$. Applying the closed operator $G^{-1}: E_o^1 \to E$ to w_n, it suffices now to prove that in E the limit $\lim_{n\to\infty} G^{-1}w_n$ exists. To this purpose, we shall consider the sequence

$$G^{-1}w_n = \sum_{\mu=o}^{n} (-1)^{\mu} G^{-1}(\varepsilon B)^{\mu} v \qquad (n \in \mathbb{N}).$$

From the relation $G^{-1}B^{\mu}v = G^{-1}(GK_1G^{-1})^{\mu}v = K_1^{\mu} G^{-1}v$, it follows that

$$G^{-1}w_n = \sum_{\mu=o}^{n} (-1)^{\mu} (\varepsilon K_1)^{\mu} G^{-1}v \qquad (n \in \mathbb{N}).$$

This sequence converges in E since we have $\varepsilon \leq \varepsilon_o < ||B||^{-1} \leq ||K_1||^{-1}$. This completes the proof.

Returning now to the two-point boundary value problem (6.1), we may ask for solutions $u(y)$, which belong to the following class:

$$\mathring{D} := \{u: [o,\delta] \to E_o^1 \mid u \in C^2((o,\delta);E) \cap C([o,\delta];E),\ u(o) = o = u(\delta)\}. \tag{6.9}$$

Clearly, in virtue of Lemma (6.E), we infer that on this function class the problem (6.1) is equivalent to

$$\left.\begin{aligned}&\tilde{A}\partial_y^2 u(y) + u(y) = \tilde{A}f(y) \quad (o < y < \delta),\\&u(o) = u(\delta) = o.\end{aligned}\right\} \qquad (6.10)$$

Here we have used the fact that $\tilde{A} = (I + \varepsilon G K_1 G^{-1})^{-1} A = (I + \varepsilon B)^{-1} A\colon E \to E_o^1$. Comparing problem (6.10) with the two-point boundary value problem (3.12), we see that we have only replaced the continuous linear operator A by the continuous linear operator $\tilde{A}$.

In view of this fact, and also in view of Corollary (6.D), we are motivated to introduce operator families $\tilde{S}_\mu(y)$ $(\mu = o,1,2)$, and $\tilde{R}_2(y)$ $(o \le y \le \delta)$ analogous to the families $S_\mu(y)$ and $R_2(y)$. To this purpose, we may replace in the definitions (3.1), (3.13) and (3.14) the reduced resolvent $R'(\lambda;A)$ by $R'(\lambda;\tilde{A})$, whereas in definition (3.29) we may substitute the operator families $S_1(y)$ and $S_2(y)$ by $\tilde{S}_1(y)$ and $\tilde{S}_2(y)$, respectively. The results of Theorems (3.L) and (3.N) apply to the new situation without any change. For later use, we shall collect them in the following theorem.

THEOREM (6.F). Let the Sector condition (H1) be satisfied, and suppose that the Root definition (H2) holds. Let the kernel function $\check{k}_1(s) := k_1(-s)$ belong to $L^1(\mathbb{R}) \cap L^\infty_{loc}(-\infty,o) \cap \check{E}$, and assume that $\varepsilon_o > o$ is any fixed number satisfying (6.8). Then we have for any $\delta \ge \delta_o$:

(i) $\tilde{R}_2(y) \in B(L^p((o,\delta);E),E)$ for each fixed $y \in [o,\delta]$ and any fixed $1 \le p \le \infty$, $\tilde{R}_2(y)$ having a bound which is independent of $\delta \ge \delta_o$. $\tilde{R}_2(\cdot)f \in C([o,\delta];E)$ and $\lim_{y\to o+} ||\tilde{R}_2(y)f|| = o = \lim_{y\to\delta-} ||\tilde{R}_2(y)f||$ for any f belonging to $L^p(o,\delta;E)$ $(1 \le p \le \infty)$.

(ii) If, in addition, $f\colon (o,\delta) \to E$ is locally uniformly Hölder continuous, and belongs to $L^p(o,\delta;E)$ for at least one $p \in [1,\infty]$, then we also have $\tilde{R}_2(\cdot)f \in C([o,\delta];E) \cap C^2((o,\delta);E)$ and $\tilde{A}\partial_y^2 \tilde{R}_2(y)f + \tilde{R}_2(y)f = \tilde{A}f(y)$ for each $y \in (o,\delta)$, where from this relation it follows that $\tilde{R}_2(y)f \in E_o^1$ for every fixed $y \in [o,\delta]$.

6.2. AN EXISTENCE AND UNIQUENESS RESULT FOR THE PROBLEM (Z)

Throughout this section we shall assume familiarity with the concepts and definitions given in Section 4.2. In particular, we shall be concerned with the remainder problem (Z) which has been stated in Theorem (4.C). Here we shall present a general existence and uniqueness theorem for solutions to (Z).

In the sequel, let $\varepsilon_o > o$ be any given number satisfying the relation (6.8). Denote by $\delta_o > o$ the constant $\delta_o = 1/\sqrt{\varepsilon_o}$, and by δ the function $\delta(\varepsilon) = 1/\sqrt{\varepsilon}$ $(o \le \varepsilon \le \varepsilon_o)$.

The way, we shall attack the problem (Z) is to transform it into an equivalent problem of homogeneous boundary data. To this purpose, we shall introduce an auxiliary function

$$W(y,t;\varepsilon) := \sum_{\mu=o}^{m} \varepsilon^{\mu}[(\delta^{-1}y-1)V_{\mu}^{1}(\delta,t) - \delta^{-1}yV_{\mu}^{o}(\delta,t)] \tag{6.11}$$

Substituting now $y = \delta x$ in (Z) and defining

$$T(y,t;\varepsilon) := Z(\delta^{-1}y,t;\varepsilon) - W(y,t;\varepsilon) \quad (o \le y \le \delta, \quad t \ge o), \tag{6.12}$$

we obtain after some rearrangements

$$\left.\begin{aligned}
&(k + \partial_t + K_c\partial_y^2 + \varepsilon K_1(k+\partial_t))T(y,t;\varepsilon) = -M_\varepsilon[W](y,t;\varepsilon) - \varepsilon^{m+1}\, w(\delta^{-1}y,t;\varepsilon)\\
&(o < y < \delta,\ t > o),\\
&T(o,t;\varepsilon) = T(\delta,t;\varepsilon) = o \quad (t \geq o),\\
&T(y,o;\varepsilon) = o \quad (o < y < \delta).
\end{aligned}\right\} \quad (T)$$

Here the function $w(x,t;\varepsilon)$ is defined by (4.19), whereas the differential relation $M_\varepsilon[W]$ can be expressed in the following form:

$$\left.\begin{aligned}
&M_\varepsilon[W](y,t;\varepsilon) = \sum_{\mu=o}^{m} \varepsilon^\mu (I+\varepsilon K_1)G^{-1}[(\delta^{-1}y-1)V_\mu^1(\delta,t) - \delta^{-1}yV_\mu^o(\delta,t)]\\
&(o \leq y \leq \delta,\quad t \geq o).
\end{aligned}\right\} \quad (6.13)$$

The advantage of the problem (T) over the original remainder problem (Z) may be seen in its homogeneous boundary data. Obviously, (T) can be identified with the two-point boundary value problem (6.1), if the solution T is required to belong to the class $\mathring{D}$ (see (6.9)). The right-hand side function $f(y)$ has to be specified by

$$\left.\begin{aligned}
f(y)(t) &:= - K_c^{-1}(M_\varepsilon[W] + \varepsilon^{m+1}w)(y,t;\varepsilon)\\
&= - \sum_{\mu=o}^{m} \varepsilon^\mu K_c^{-1}(I+\varepsilon K_1)G^{-1}[(\delta^{-1}y-1)V_\mu^1(\delta,t) - \delta^{-1}yV_\mu^o(\delta,t)]\\
&- \varepsilon^{m+1}[(\partial_x^2 + K_c^{-1}K_1G^{-1})\, U_m\ (x = \delta^{-1}y,t)\\
&+ K_c^{-1}K_1G^{-1}(V_m^o(y,t) + V_m^1(\delta-y,t))] \quad (o < y < \delta,\ t > o).
\end{aligned}\right\} \quad (6.14)$$

Now it is clear that we want to apply Theorem (6.F) in order to obtain an existence and uniqueness result for the solutions to (T). The only crucial condition in Theorem (6.F) is the condition of Hölder continuity of the function $f(y)$. In order to guarantee the proper continuity of $f(y)$, we shall assume that the following hypothesis holds.

$(H)_{m+1,\alpha}$ REGULARITY CONDITION. There exists a number $\alpha \in (o,1)$ such that r belongs to the Hölder class $C^{2m+\alpha}([o,1];E)$ and such that h belongs to the Hölder class $C^{2m+2+\alpha}[o,1]$.

The following result on existence and uniqueness is central.

THEOREM (6.G). Let the Sector condition (H1) and the Compatibility condition (H3) be satisfied, and suppose that the Root definition (H2) and the Regularity condition $(H)_{m+1,\alpha}$ hold. Furthermore, let the kernel function $\check{k}_1(s) := k_1(-s)$ belong to $L^1(\mathbb{R}) \cap L^\infty_{loc}(-\infty,o) \cap \check{E}$, and assume that $\varepsilon_o > o$ is any fixed number satisfying (6.8). Then, for any $o < \varepsilon \le \varepsilon_o$, the initial boundary value problem (Z) has a unique solution $Z \in C^2((o,1);E) \cap C([o,1];E)$ satisfying $Z(x,\cdot;\varepsilon) \in E^1_o$ for every $x \in [o,1]$. This solution admits a representation in the form

$$Z(x,t;\varepsilon) = T(\delta x,t;\varepsilon) + W(\delta x,t;\varepsilon) \qquad (o \le x \le 1,\ t \ge o),$$

where $\delta := 1/\sqrt{\varepsilon}$, and where W is defined by (6.11). The function T is given by $T(y,t;\varepsilon) = (\tilde{R}_2(y)f)(t;\varepsilon)$ $(o \le y \le \delta,\ t \ge o)$, where f is defined by (6.14).

<u>Proof.</u> Having established the Höder continuity of the function f, then, in view of the preceding considerations, the existence part of the proof is

immediately obtained from Theorem (6.F). Thus, we shall prove the Hölder continuity of f. Considering each term in formula (6.14) individually we shall start with the sum

$$\sum_{\mu=o}^{m} \varepsilon^{\mu} K_c^{-1}(I + \varepsilon K_1)G^{-1}[(\delta^{-1}y-1)V_{\mu}^{1}(\delta,t) - \delta^{-1}yV_{\mu}^{o}(\delta,t)].$$

Since this is a continuous linear function of y varying on a bounded interval $[o,\delta]$, we have nothing to prove.

Next, we shall consider the term $(\partial_x^2 + K_c^{-1}K_1G^{-1})\, U_m\, (x = \delta^{-1}y,t)$. Using the relation (4.6) together with the representation formula (4.8), we will see that

$$\left.\begin{aligned} -(\partial_x^2 + K_c^{-1}K_1G^{-1})U_m(x,t) &= K_c^{-1}G^{-1}U_{m+1}(x,t) \\ &= (-1)^m \sum_{\nu=o}^{m} K_c^{-1}G^{-1}B_{m+1,\nu}[(\Delta^{\nu}r)(x,t) - (K_cg)(t)\cdot(\Delta^{\nu+1}h)(x)]. \end{aligned}\right\} \quad (6.15)$$

Replacing here the variable x by $\delta^{-1}y$, and observing the hypothesis $(H)_{m+1,\alpha}$, we easily deduce the Hölder continuity.

Finally, considering the last term in (6.14), we infer by (5.24) that

$$K_c^{-1}K_1G^{-1}[V_m^o(y,t) + V_m^1(\delta-y,t)] = f_{m+1}^o(y)(t) + f_{m+1}^1(\delta-y)(t). \qquad (6.16)$$

Thus, the required Hölder continuity is guaranteed by Corollary (5.K). This completes the existence proof. The uniqueness follows by a standard procedure.

6.3. THE UNIFORM BOUNDEDNESS OF THE REMAINDER FUNCTION

We are able now to prove the following singular perturbation theorem

THEOREM (6.H). Let the Sector condition (H1) and the Compatibility conditions (H3) be satisfied, and suppose that the Root definition (H2) and the Regularity condition $(H)_{m+1,\alpha}$ hold. Furthermore, let the kernel function $\check{k}_1(s) := k_1(-s)$ belong to $L^1(\mathbb{R}) \cap L^\infty_{loc}(-\infty,o) \cap \check{E}$, and assume that $\varepsilon_o > o$ is any fixed number satisfying the condition (6.8). Then for any $o < \varepsilon \le \varepsilon_o$, the initial boundary value problem (P_ε) admits a unique solution $u \in C^2((o,1);E) \cap C([o,1];E)$ which can be represented by

$$u(x,t;\varepsilon) = \bar{u}(x,t;\varepsilon) + Z(x,t;\varepsilon) \qquad (o \le x \le 1,\ t \ge o).$$

Here $\bar{u}(x,t;\varepsilon) = U(x,t;\varepsilon) + V^o(x,t;\varepsilon) + V^1(x,t;\varepsilon)$ is an m-th order approximate solution to (P_ε), where the functions U, V^o and V^1 are defined by (4.1) and (4.15) respectively. In particular, the functions V^j $(j = o,1)$ are boundary layer terms having the properties, that for any fixed $\eta \in (o,1)$

$$\max_{x\in[\eta,1]} ||V^o(x,\cdot;\varepsilon)||_E \to o, \quad \max_{x\in[o,1-\eta]} ||V^1(x,\cdot;\varepsilon)||_E \to o \quad \text{as} \quad \varepsilon \to o+, \tag{6.17}$$

with an exponential rate of decay. Moreover, the remainder function Z is small of order ε^{m+1} in the sense of

$$\max_{x\in[o,1]} ||Z(x,\cdot;\varepsilon)||_E = O(\varepsilon^{m+1}) \quad \text{as} \quad \varepsilon \to o+. \tag{6.18}$$

Obviously, the proof of Theorem (6.H) reduces to the verfication of the relations (6.17) and (6.18). Since the proof of these relations is more involved, we shall divide it into several steps.

First of all we shall be concerned with the boundary layer terms

$$V^j(x,t;\varepsilon) = \sum_{\mu=o}^{m} \varepsilon^{\mu} V_{\mu}^j(\tilde{x}_j,t) \quad (o \le x \le 1,\ t \ge o,\ \ j = o,1) \tag{6.19}$$

where the variable $\tilde{x}_j$ has to be substituted either by $\tilde{x}_o = x/\sqrt{\varepsilon}$ or by $\tilde{x}_1 = (1-x)/\sqrt{\varepsilon}$. The following proposition precedes the main argument.

PROPOSITION (6.I). Under the hypotheses of Theorem (6.H), for any fixed $y_o > o$, there exist constants $C > o$ and $o < \beta < \frac{1}{2}\sqrt{\rho}\sin\theta/2$ such that

$$||V_{\mu}^j(y,\cdot)|| \le Ce^{-\beta y} \quad \text{for every} \quad y \ge y_o \quad (\mu = o,1,\ldots,m;\ j = o,1) \tag{6.20}$$

<u>Proof.</u> Let us first consider the case where $\mu=o$. An application of Theorem (5.A) shows that $V_o^j(y,t) = S_o(y)\,[g_j(t) - h(j)e^{-kt}]$ $(j = o,1)$, and hence the property $v_j(t) := g_j(t) - h(j)e^{-kt} \in E_o^1$ together with Proposition (3.B) imply that

$$||V_o^j(y,\cdot)|| \le C||v_j||(1+1/y_o)e^{-y\sqrt{\rho}\sin\theta/2} \quad (y \ge y_o) \tag{6.21}$$

Thus the proposition is proved for $\mu=o$.

In the general case, where $o < \mu \le m$, we obtain from Theorem (5.I) the representation $V_{\mu}^j(y,t) = -\,S_o(y)U_{\mu}(j,t) + (R_1(y)f_{\mu}^j)(t)$ $(j = o,1)$. By hypothesis, we have $U_{\mu}(j,\cdot) \in E_o^1$, and hence we deduce by a repetition of the foregoing arguments that

$$\|S_o(y)U_\mu(j,\cdot)\| \le C\|U_\mu(j,\cdot)\|\,(1+1/y_o)e^{-y\sqrt{\rho}\,\sin\theta/2} \qquad (y \ge y_o). \tag{6.22}$$

It remains now to find an appropriate estimate for the terms $(R_1(y)f_\mu^j)(t)$, where the functions f_μ^j are given by (5.24), and where $R_1(y)$ is defined by (3.23). After some rearrangements, we will see that (3.23) can be equivalently written in the following form

$$\left.\begin{aligned} R_1(y)f_\mu^j = \frac{1}{2}\{ &\int_{y/2}^{\infty} S_1(z)f_\mu^j(y+z)dz + \int_{y/2}^{y} S_1(z)f_\mu^j(y-z)dz - \\ &\int_o^\infty S_1(y+z)f_\mu^j(z)dz\} \\ &+ \frac{1}{2}\int_o^{y/2} S_1(z)[f_\mu^j(y+z) + f_\mu^j(y-z) - 2f_\mu^j(\tfrac{y}{2})]dz \\ &+ f_\mu^j(\tfrac{y}{2})\int_o^{y/2} S_1(z)dz. \end{aligned}\right\} \tag{6.23}$$

From Proposition (3.B), we obtain

$$\|S_1(z)\| \le C\, e^{-z\sqrt{\rho}\,\sin\theta/2} \qquad (z \ge o). \tag{6.24}$$

Combining this with the property $f_\mu^j \in L^1(\mathbb{R}_+;E)$ (see Corollary (5.K)), we conclude that in formula (6.23) the term within the waved brackets can be bounded by

$$C\|f_\mu^j\|_{L^1(\mathbb{R}_+;E)}\; e^{-y\sqrt{\rho}/2\,\sin\theta/2} \qquad (y \ge y_o,\ j = o,1). \tag{6.25}$$

Next, an application of Corollary (5.K) yields

$$||f_\mu^j(y+z) + f_\mu^j(y-z) - 2f_\mu^j(\tfrac{y}{2})|| \le C(|z+\tfrac{y}{2}|^\alpha + |z-\tfrac{y}{2}|^\alpha)(e^{-\beta y} + y^{-\alpha} \cdot e^{-\beta y/2})$$

$(o < z < y/2, \quad o < \alpha < 1)$.

Consequently, by (6.24)

$$\left.\begin{aligned} &||\int_o^{y/2} S_1(z)[f_\mu^j(y+z) + f_\mu^j(y-z) - 2f_\mu^j(\tfrac{y}{2})]\, dz\, || \\ &\le Cy^\alpha e^{-y\sqrt{\rho}/2 \sin\theta/2} (e^{-\beta y} + y^{-\alpha} \cdot e^{-\beta y/2}) \le Ce^{-\beta y} \quad (y \ge y_o). \end{aligned}\right\} \quad (6.26)$$

Thus it remains to determine an estimate for $||f_\mu^j(\frac{y}{2})||$. For $\mu=1$, we infer from (5.6) and (6.21) that

$$\left.\begin{aligned} &||f_1^j(\tfrac{y}{2})|| = ||K_c^{-1}K_1K_cS_o(\tfrac{y}{2})A^{-1}v_j|| \\ &\le C||A^{-1}v_j||\,(1+1/y_o)e^{-y\sqrt{\rho}\,\sin\theta/2} \quad (y \ge y_o) \end{aligned}\right\} \quad (6.27)$$

In case of $\mu=2$, we obtain from formula (5.19):

$$||f_2^j(\tfrac{y}{2})|| \le$$

$$\le ||K_c^{-1}K_1K_c||\,\{||S_o(\tfrac{y}{2})A^{-1}U_1(j,\cdot)|| + ||f_1^j(\tfrac{y}{2})|| + \frac{1}{2}\sum_{\nu=1}^{4} ||I_\nu(\tfrac{y}{2})||\}.$$

Applying now Proposition (5.G) and Corollary (5.H), and using the estimates (6.22) and (6.27), it follows that

$$||f_\mu^j(\tfrac{y}{2})|| \le C\, e^{-\beta y} \quad (y \ge y_o, \quad j = o,1;\ \mu = 2). \quad (6.28)$$

By an induction argument, it is now easy to see that the estimate (6.28) also holds für $2 \le \mu \le m$.

Putting the estimates (6.25, 6.26 and 6.28) into formula (6.23), we conclude that

$$||R_1(y)f_\mu^j|| \le C\ e^{-\beta y} \quad (y \ge y_0,\ \ j = 0,1). \tag{6.29}$$

This completes the proof of Proposition (6.I).

Proposition (6.I) yields the following result on the exponential decay of the boundary layer terms.

LEMMA (6.K). Under the hypotheses of Theorem (6.H), for any fixed $\eta \in (0,1)$, there exist constants $C > 0$ and $0 < \beta < \frac{1}{2}\sqrt{\rho}\sin\theta/2$ such that for $\varepsilon \to 0+$:

$$\left.\begin{aligned} \max_{\eta\le x\le 1} ||V^0(x,\cdot;\varepsilon)|| &\le C\ e^{-\beta\eta/\sqrt{\varepsilon}}, \\ \max_{0\le x\le 1-\eta} ||V^1(x,\cdot;\varepsilon)|| &\le C\ e^{-\beta\eta/\sqrt{\varepsilon}} \end{aligned}\right\} \tag{6.30}$$

<u>Proof.</u> We shall consider V^0 and V^1 separately. Starting with the boundary layer term V^0, in formula (6.19), we substitute $\tilde{x}_0 = x/\sqrt{\varepsilon}$ by y. Letting $y_0 = \eta/\sqrt{\varepsilon_0}$, it is obvious that $y \ge y_0$ for all $0 < \varepsilon \le \varepsilon_0$ and all $\eta \le x \le 1$. Hence, we may apply Proposition (6.I) in order to see that

$$\max_{\eta\le x\le 1} ||V_\mu^0(\tilde{x}_0,\cdot)|| \le C\ e^{-\beta\eta/\sqrt{\varepsilon}} \quad (0 < \varepsilon \le \varepsilon_0).$$

This yields for any $o < \varepsilon \le \varepsilon_o$:

$$\max_{\eta \le x \le 1} \|V^o(x,\cdot;\varepsilon)\| \le C\, e^{-\beta\eta/\sqrt{\varepsilon}} \sum_{\mu=o}^{m} \varepsilon^{\mu} \le \frac{1-\varepsilon_o^{m+1}}{1-\varepsilon_o}\, C\, e^{-\beta\eta/\sqrt{\varepsilon}}$$

as claimed.

Next, we consider the boundary layer term V^1. Letting now $y = (1-x)/\sqrt{\varepsilon}$, and defining y_o as before, it is immediately obvious that $y \ge y_o$ for all $o < \varepsilon \le \varepsilon_o$ and all $o \le x \le \eta - 1$. Thus, we may repeat the foregoing arguments in order to verify the second relation in (6.30). This completes the proof of Lemma (6.K).

Clearly, the relations (6.30) imply the statements (6.17) of Theorem (6.H). To complete the proof of Theorem (6.H), we shall be concerned with the estimate (6.18) of the remainder function. Recall that

$$Z(x,t;\varepsilon) = T(x/\sqrt{\varepsilon},t;\varepsilon) + W(x/\sqrt{\varepsilon},t;\varepsilon), \tag{6.31}$$

where $W(x/\sqrt{\varepsilon},t;\varepsilon) = \sum_{\mu=o}^{m} \varepsilon^{\mu}\,[(x-1)V_{\mu}^{1}(1/\sqrt{\varepsilon},t) - xV_{\mu}^{o}(1/\sqrt{\varepsilon},t)]$. By Proposition (6.I) this yields

$$\max_{o \le x \le 1} \|W(x/\sqrt{\varepsilon},\cdot;\varepsilon)\| \le \frac{1-\varepsilon_o^{m+1}}{1-\varepsilon_o}\, C\, e^{-\beta/\sqrt{\varepsilon}} \qquad (o < \varepsilon \le \varepsilon_o),$$

where $C > o$ and $o < \beta < \frac{1}{2}\sqrt{\rho}\,\sin\theta/2$ are constants not depending upon ε. Clearly it also follows that $\max_{o \le x \le 1} \|W(x/\sqrt{\varepsilon},\cdot;\varepsilon)\| = O(\varepsilon^{\gamma})$ as $\varepsilon \to o+$, for any power $\gamma > o$. Therefore, it remains to find an appropriate estimate for the function $T(x/\sqrt{\varepsilon},t;\varepsilon) = (\tilde{R}_2(x/\sqrt{\varepsilon})f)(t;\varepsilon)$. Here, $f(y)(t)$ is given by formula (6.14). In order to find the proper estimate, we shall proceed in two steps. First we shall be concerned with the function $\tilde{R}_2(x/\sqrt{\varepsilon})M_\varepsilon[W]$, where

$$M_\varepsilon[W] = \sum_{\mu=o}^{m} \varepsilon^\mu K_c^{-1}(I+\varepsilon K_1)G^{-1}[(x-1)V_\mu^1(1/\sqrt{\varepsilon},t) - xV_\mu^o(1/\sqrt{\varepsilon},t)]. \qquad (6.32)$$

LEMMA (6.L). Under the hypotheses of Theorem (6.H), we have

$$\max_{o\leq x\leq 1} ||\tilde{R}_2(x/\sqrt{\varepsilon})\, M_\varepsilon[W]|| = O(\varepsilon^\gamma) \quad \text{as} \quad \varepsilon \to o+,$$

where $\gamma>o$ is an arbitrary power.

<u>Proof.</u> Letting $\delta := 1/\sqrt{\varepsilon}$ and $y := \delta x$ as before, by definition (3.29), we will see that

$$||\tilde{R}_2(y)M_\varepsilon[W]|| \leq \frac{1}{2} \max_{o\leq z\leq \delta} ||M_\varepsilon[W](z)||$$
$$\times \{ \int_o^\delta [||\tilde{S}_2(z+\zeta)|| + ||\tilde{S}_2(2\delta+z-\zeta)|| + ||\tilde{S}_2(2\delta-z-\zeta)|| + ||\tilde{S}_2(2\delta-z+\zeta)||]d\zeta$$
$$+ \int_o^z ||\tilde{S}_1(z-\zeta)||d\zeta + \int_z^\delta ||\tilde{S}_1(\zeta-z)||\, d\zeta\}.$$

Since

$$||\tilde{S}_1(z)|| \leq C\, e^{-z\sqrt{\rho}\, \sin\theta/2},$$
$$||\tilde{S}_2(z)|| \leq (1 - e^{-2\delta_o\sqrt{\rho}\, \sin\theta/2})^{-1}\, ||\tilde{S}_1(z)|| \qquad (6.33)$$

for all $z \geq o$ and uniformly with respect to $\delta \geq \delta_o = 1/\sqrt{\varepsilon_o}$, it is immediately obvious that

$$||\tilde{R}_2(y)M_\varepsilon[W]|| \leq C \max_{0\leq z\leq\delta} ||M_\varepsilon[W](z)|| \quad (o \leq y \leq \delta), \tag{6.34}$$

where $C > o$ does not depend upon $\delta \geq \delta_o$. Now, from definition (6.32) it follows that

$$\left.\begin{aligned}&\max_{o\leq z\leq\delta} ||M_\varepsilon[W](z)|| \leq ||K_c^{-1}|| \,(1+\varepsilon_o\, ||K_1||)\\ &\times \sum_{\mu=o}^{m} \varepsilon^\mu \,(||G^{-1}V_\mu^1(1/\sqrt{\varepsilon},\cdot)|| + ||G^{-1}V_\mu^o(1/\sqrt{\varepsilon},\cdot)||).\end{aligned}\right\} \tag{6.35}$$

Furthermore, in virtue of formula (5.24), we have $f_\mu^j(1/\sqrt{\varepsilon}) = - K_c^{-1}K_1G^{-1}V_{\mu-1}^j(1/\sqrt{\varepsilon},\cdot)$ $(\mu = 1,\dots,m+1;\ j = o,1)$, and hence we may adapt the estimates (6.27 and 6.28) to the present situation in order to obtain $||G^{-1}V_\mu^j(1/\sqrt{\varepsilon},\cdot)|| \leq C\, e^{-\beta/\sqrt{\varepsilon}}$ $(\mu = o,\dots,m;\ j = o,1)$, where $C>o$ and $o < \beta < \sqrt{\rho}\,\sin\theta/2$ are constants not depending upon ε. A combination of this inequality with the estimates (6.33) and (6.34) yields

$$\max_{0\leq x\leq 1} ||\tilde{R}_2(x/\sqrt{\varepsilon})M_\varepsilon[W]|| \leq \frac{1-\varepsilon^{m+1}}{1-\varepsilon}\, C\, e^{-\beta/\sqrt{\varepsilon}} \quad (o < \varepsilon \leq \varepsilon_o),$$

from which the assertion is easily obtained. This completes the proof.

In the second step we shall be concerned with the function

$$\left.\begin{aligned}&\varepsilon^{m+1}\, w(x,t;\varepsilon) = \varepsilon^{m+1}\, [(\partial_x^2 + K_c^{-1}K_1G^{-1})U_m(x,t)\\ &+ K_c^{-1}K_1G^{-1}(V_m^o(x/\sqrt{\varepsilon},t) + V_m^1((1-x)/\sqrt{\varepsilon},t))].\end{aligned}\right\} \tag{6.36}$$

LEMMA (6.M). Under the hypotheses of Theorem (6.H), we have $\max_{0\leq x\leq 1} ||\tilde{R}_2(x/\sqrt{\varepsilon})w|| = O(1)$ as $\varepsilon \to o+$.

Proof. In this proof, we shall use two different arguments according to the different members in formula (6.36), namely

$w_1(x,t) = (\partial_x^2 + K_c^{-1}K_1G^{-1})U_m(x,t)$ and

$w_2(x,t;\varepsilon) = K_c^{-1}K_1G^{-1}[V_m^o(x/\sqrt{\varepsilon},t) + V_m^1((1-x)/\sqrt{\varepsilon},t)]$. Starting with w_1 and using the hypothesis $(H)_{m+1,\alpha}$, one obtains from the identity (6.15)

$$\max_{o\le x\le 1} ||w_1(x,\cdot)|| = \max_{o\le x\le 1} ||K_c^{-1}G^{-1}U_{m+1}(x,\cdot)|| = O(1).$$

Using now an estimate analogous to (6.34), this yields $\max_{o\le x\le 1} ||\tilde{R}_2(x/\sqrt{\varepsilon})w_1|| = O(1)$. Next, we shall be concerned with the function w_2 which is, by formula (6.16), equivalent to

$w_2(x,t;\varepsilon) = -[f_{m+1}^o(x/\sqrt{\varepsilon})](t) - [f_{m+1}^1((1-x)/\sqrt{\varepsilon})](t)$. An application of Corollary (5.K) shows that $f_{m+1}^j \in L^1(\mathbb{R}_+;E)$ $(j = o,1)$. Hence, letting $y := x/\sqrt{\varepsilon}$, we infer from the definition of $\tilde{R}_2$ that

$$||\tilde{R}_2(y)f_{m+1}^j|| \le \frac{1}{2}\int_o^\infty ||f_{m+1}^j(\zeta)||d\zeta$$

$$\times \{\max_{o\le z\le\delta} [||\tilde{S}_2(y+z)|| + ||\tilde{S}_2(2\delta+y-z)|| + ||\tilde{S}_2(2\delta-y-z)|| + ||\tilde{S}_2(2\delta-y+z)||]$$

$$+ \max_{o\le z\le y} ||\tilde{S}_1(y-z)|| + \max_{y\le z\le\delta} ||\tilde{S}_1(z-y)||\}. \quad (o \le y \le \delta).$$

Using the estimates (6.33), this yields $\max_{o\le x\le 1} ||\tilde{R}_2(x/\sqrt{\varepsilon})f_{m+1}^j|| = O(1)$ $(j = o,1)$, which completes the proof of Lemma (6.M).

To conclude the proof of Theorem (6.H), we observe that in virtue of Lemma (6.M) it follows that

$$\max_{o\le x\le 1} ||\tilde{R}_2(x/\sqrt{\varepsilon})\varepsilon^{m+1}w|| = O(\varepsilon^{m+1}) \quad \text{as} \quad \varepsilon \to o+.$$

Therefore, the desired relation (6.18) is easily obtained by combining this result with the result of Lemma (6.L).

Bibliography

1 A. Ahiezer, N. Ahiezer, G. Lyubarskii, Effective boundary conditions on the surface of separation between multiplying and moderating media. Ž. Techn. Fiz. 27 (1957), 822-829 = Sov. Phys.-Techn. Phys. 2 (1957/58), 749-757.

2 A.V. Balakrishnan, Fractional powers of closed operators and the semigroups generated by them. Pacific J. Math. 10 (1960), 419-437.

3 V. Barbu, Nonlinear semigroups and differential equations in Banach spaces. (Bucureşti, România: Editura Academiei, 1976).

4 H. Brézis, Opérateurs maximaux monotones et semigroupes de contractions dans les espaces de Hilbert. Math. Studies 5. (Amsterdam-London: North-Holland Publ. Comp., 1973).

5 F.E. Browder, Fixed points theorems for nonlinear semicontractive mappings in Banach spaces. Arch. Rat. Mech. Anal. 21 (1966), 259-269.

6 F.E. Browder, Nonlinear accretive operators in Banach spaces. Bull. Amer. Math. Soc. 73 (1967), 470-476.

7 F.E. Browder, Semicontractive and semiaccretive nonlinear mappings in Banach spaces. Bull. Amer. Math. Soc. 74 (1968), 660-665.

8 R.W. Carroll, Abstract methods in partial differential equations. (New York: Harper and Row, 1969).

9 J. Cole, Perturbation techniques in applied mathematics. (Waltham, Mass.: Blaisdell, 1968).

10 C.M. Dafermos, Asymptotic stability in viscoelasticity. Arch. Rat. Mech. Anal. 37 (1970), 297-308.

11 C.M. Dafermos, An abstract Volterra equation with applications to linear viscoelasticity. J. Diff. Equs. 7 (1970), 554-569.

12 C.M. Dafermos, Contraction semigroups and trends to equilibrium in continuum mechanics. Applications of methods of functional analysis to problems in mechanics. Eds. P. Germain and B. Nayroles. Springer Lecture Notes 503, 295-306. (Berlin-Heidelberg-New York: Springer, 1976).

13 N. Dunford and J.T. Schwartz, Linear operators. I. General theory. (New York: Interscience Publ., 1958).

14 M. Van Dyke, Perturbation methods in fluid mechanics. (New York: Academic Press, 1964).

15 W. Eckhaus, Matched asymptotic expansions and singular perturbations. Math. Studies 6. (Amsterdam-London: North-Holland Publ. Comp., 1973).

16 W. Eckhaus and E.M. De Jager, Asymptotic solutions of singular perturbation problems for linear differential equations of elliptic type. Arch. Rat. Mech. Anal. 23 (1966), 26-86.

17 A. Erdélyi, Singular perturbations. Trends in applications of pure mathematics to mechanics. Ed. G. Fichera, 53-62. (London: Pitman Publ., 1976).

18 L.E. Fraenkel, On the method of matched asymptotic expansions. I. A matching principle; II. Some applications of the composite series; III. Two boundary value problems. Proc. Cambridge Phil. Soc. 65 (1969), 209-284.

19 K.O. Friedrichs, Symmetric positive linear differential equations. Comm. Pure Appl. Math. 11 (1958), 333-418.

20 E. Gerlach, Zur Theorie einer Klasse von Integrodifferentialgleichungen. (Technische Universität Berlin, Dissertation, 1969).

21 H. Grabmüller, Über die Lösbarkeit einer Integrodifferentialgleichung aus der Theorie der Wärmeleitung. Methoden und Verfahren der Mathematischen Physik. 10. Eds. B. Brosowski and E. Martensen, 117-137. (Mannheim: Bibl. Institut 1973).

22 H. Grabmüller, On linear theory of heat conduction in materials with memory. Existence and uniqueness theorems for the final value problem. Proc. Royal Soc. Edinburgh 76A (1976), 119-137.

23 H. Grabmüller, On linear partial integro differential equations with a small parameter. Proc. of the Conference on Ordinary and Partial Differential Equations. Eds. W.N. Everitt and B.D. Sleeman. Springer Lecture Notes 564, 125-134. (Berlin-Heidelberg-New York: Springer, 1976).

24 M.E. Gurtin and A.C. Pipkin, A general theory of heat conduction with finite ware speeds. Arch. Rat. Mech. Anal. 31 (1968), 113-126.

25 A. Van Harten, Singularly perturbed non-linear 2nd order elliptic boundary value problems. (Utrecht, Thesis, 1975).

26 E. Hille and R.S. Phillips, Functional analysis and semi-groups. (Providence, R.I.: Amer. Math. Soc., 1957).

27 F. Hoppensteadt, On quasilinear parabolic equations with a small parameter. Comm. Pure Appl. Math. 24 (1971), 17-38.

28 R.P. Kanwal, Applications of the technique of matched asymptotic expansions to various fields of mathematical physics. Proc. Symp. Analyt. Meth. in Math. Phys., 479-485. (Bloomington, Indiana: Indiana Univ. Press, 1969).

29 S. Kaplun, Fluid mechanics and singular perturbations. Eds. P.A. Lagerstrom, L.N. Howard and C.S. Liu. (New York: Academic Press, 1967).

30 S. Kaplun and P.A. Lagerstrom, Asymptotic expansions of Navier-Stokes solutions for small Reynolds numbers. J. Math. Mech. 6 (1957), 585-593.

31 T. Kato, Perturbation theory of linear operators. (Berlin-Heidelberg-New York: Springer, 1966).

32 T. Kato, Semigroups and temporarily inhomogeneous evolution equations. Equazione Differenziale Astratte. (C.I.M.E. Edizione Cremonese, Roma, 1963).

33 T. Kato, Accretive operators and nonlinear evolution equations in Banach spaces. Nonlinear functional analysis. Proc. Symp. Pure Math. Vol. 18, Part. I. Ed. F.E. Browder, 138-161. (Providence, R.I.: Amer. Math. Soc., 1970).

34 M.G. Krein, Integral equations on a half line with kernel depending upon the difference of the arguments. Uspehi Mat. Nauk (N.S.) 13 (1958), 3-120 = Amer. Math. Soc. Transl. Ser. 2., 22 (1962), 163-288.

35 P.A. Lagerstrom, Some recent developments in the theory of singular perturbations. Problems in analysis. A symposium in honour of S. Bochner. 261-271. (New Jersey: Princeton Univ. Press, 1970).

36 J. Meixner, On the linear theory of heat conduction. Arch. Rat. Mech. Anal. 39 (1970), 108-130.

37 R.K. Miller, An integrodifferential equation for rigid heat conductors with memory, in preparation.

38 J.W. Nunziato, On heat conduction in materials with memory. Quart. Appl. Math. 29 (1971), 187-204.

39 R.S. Phillips, Dissipative hyperbolic systems. Trans. Amer. Math. Soc. 86 (1957), 109-173.

40 R.S. Phillips, Dissipative operators and hyperbolic systems of partial differential equations. Trans. Amer. Math. Soc. 90 (1959), 193-254.

41 L. Prandtl, Über Flüssigkeiten bei sehr kleiner Reibung. Verh. III. Internat. Math. Kongr., 484-491 (Leipzig: Teubner, 1905).

42 I.M. Rapoport, On a class of singular integral equations. Dokl. Akad. Nauk. SSSR 59 (1948), 1403-1406.

43 M. Slemrod, Existence, uniqueness, stability for a simple fluid with fading memory. Bull. Amer. Math. Soc. 82 (1976), 581-583.

44 A.E. Taylor, Introduction to functional analysis. (New York: John Wiley, 1958).

45 M.M. Vainberg, Variational method and method of monotone operators in the theory of nonlinear equations. A Halsted Press Book. (New York - Toronto: John Wiley, 1973).

46 M.M. Vainberg, On the convergence of a steepest-descent method for non-linear equations. Dokl. Akad. Nauk SSSR 130 (1960).

47 N.P. Vekua, Linear integro-differential equations with small parameters for higher derivatives. Problems of continuum mechanics. SIAM (1961), 592-601.

48 M.I. Višik and L.A. Lyusternik, Regular degeneration and boundary layer for linear differential equations with small parameter. Uspehi Mat. Nauk 12 (1957), 3-122 = Amer. Math. Soc. Transl. Ser. 2, 20 (1960), 239-364.

49 W. Walter, Differential and integral inequalities. (Berlin-Heidelberg-New York: Springer, 1970).

50 N. Wiener and E. Hopf, Über eine Klasse singulärer Integralgleichungen. Sitz. Ber. Preuss. Akad. Wiss. Berlin. Phys.-Math. Kl. 30/32 (1931), 696-706.